SPRINGER SERIES
IN PERCEPTION ENGINEERING

Series Editor: Ramesh C. Jain

Springer Series
in Perception Engineering

Kurt D. Skifstad

High-Speed Range Estimation Based on Intensity Gradient Analysis

With 108 Illustrations

Springer-Verlag
New York Berlin Heidelberg London
Paris Tokyo Hong Kong Barcelona

Kurt D. Skifstad
Artificial Intelligence Laboratory
The University of Michigan
1101 Beal Avenue
Ann Arbor, MI 48109 USA

Series Editor

Ramesh C. Jain
Electrical Engineering
and Computer Science Department
The University of Michigan
Ann Arbor, MI 48109 USA

Library of Congress Cataloging-in-Publication Data
Skifstad, Kurt D.
   High-speed range estimation based on intensity gradient analysis :
Kurt D. Skifstad.
      p.   cm. — (Springer series in perception engineering)
   Includes bibliographical references (p.   ) and index.
   1. Image processing.   I. Title: Intensity gradient
analysis. III. Series.
TA1632.S624   1991
621.36'7—dc20   90-19775
                                                      CIP

Printed on acid-free paper.

Camera-ready copy provided by the author.

9 8 7 6 5 4 3 2 1

ISBN-13: 978-1-4612-7801-6   e-ISBN-13: 978-1-4612-3112-7
DOI: 10.1007/978-1-4612-3112-7

Preface

A fast and reasonably accurate perception of the environment is essential for successful navigation of an autonomous agent. Although many modes of sensing are applicable to this task and have been used, vision remains the most appealing due to its passive nature, good range, and resolution. Most vision techniques to recover depth for navigation use stereo. In the last few years, researchers have started studying techniques to combine stereo with the motion of the camera. Skifstad's dissertation proposes a new approach to recover depth information using known camera motion. This approach results in a robust technique for fast estimation of distances to objects in an image using only one translating camera.

A very interesting aspect of the approach pursued by Skifstad is the method used to bypass the most difficult and computationally expensive step in using stereo or similar approaches for the vision-based depth estimation. The correspondence problem has been the focus of research in most stereo approaches. Skifstad trades the correspondence problem for the known translational motion by using the fact that it is easier to detect single pixel disparities in a sequence of images rather than arbitrary disparities after two frames. A very attractive feature of this approach is that the computations required to detect single pixel disparities are local and hence can be easily parallelized. Another useful feature of the approach, particularly in navigation applications, is that the closer objects are detected earlier.

Skifstad's work demonstrates how a seemingly difficult problem in machine vision can be simplified by considering information that can be obtained from non-vision sources. Many researchers are now realizing the importance of this idea and are developing approaches called *active vision,* or *purposive vision*. Skifstad argues, in the spirit of Neisser, that all vision should be purposive and active and hence should utilize any and all relevant information that may help in a given task.

Skifstad considers his approach an engineering approach and goes on to study the performance of his approach in different situations by performing several carefully designed experiments. He discusses strengths and limitations of the approach by demonstrating its performance under varying situations. Skifstad demonstrates how engineering concepts can be used to simplify pragmatic machine perception tasks. The proposed algorithm emphasizes what is needed to solve the navigation problem, how the cameras work, and how to estimate depths using practical cameras, common hardware, and computers. His emphasis on the task results in an algorithm that is robust and computationally very efficient.

I believe that the algorithm and the engineering approach presented in this book will be applicable in many other areas of machine perception.

Ramesh Jain

Ann Arbor, Michigan

Contents

Acknowledgements

The work described in this book was submitted as partial fulfillment of the requirements for the degree Doctor of Philosophy of Computer Science and Engineering at the Horace H. Rackham School of Graduate Studies, The University of Michigan, Ann Arbor, Michigan.

I would like to thank my advisor, Prof. Ramesh Jain, for all the support, guidance, advice, freedom and friendship he has given me over the past four years. For this, I am truly grateful and deeply indebted. Thank you, Ramesh.

To Dr. Paul Besl, Prof. Trevor Mudge, Prof. David Wehe, and Prof. Terry Weymouth (the other members of my doctoral committee): thank you for the time invested reading and making comments on my dissertation.

To the faculty, staff, and students at the University of Michigan Artificial Intelligence Laboratory, thanks for making it a great place to do research.

I would like to thank the Department of Energy for supporting my research (Department of Energy Grant DE-FG02-86NE37969). I would also like to thank the DoE Robotics Group at the University of Michigan for their support and feedback (and pizza, too!).

To all my friends who were there through all the crazy times, helped me celebrate the good times, made me keep it all in perspective, and, most importantly, helped me keep my sanity and sense of humor: thanks guys.

To my parents, Dr. James G. Skifstad and Patricia Skifstad, who have challenged and supported me from day one: thanks for being there.

And, finally, thanks to the birds on the beach in Kailua Bay, who made me wonder what this depth perception thing was all about anyway ...

1

Introduction

For an autonomous agent to operate in an unknown environment, it must be able to perceive quickly and accurately the locations of the objects in that environment. This task must be accomplished quickly because the agent must be able to respond to events in its environment in a timely fashion. Accuracy is, of course, important because an agent may not be able to reason correctly about its environment with inaccurate data.

Although many modes of sensing are applicable to this task, vision is the most appealing: it has virtually infinite range, excellent spatial resolution, and is passive. It is no accident that the evolutionary process has selected vision as the primary mode of sensing used to accomplish this task. In fact, the efficacy of visual sensors in biological "agents" is a strong argument for the use of visual sensors in man-made, or artificial agents.

Attempts to endow artificial agents with vision-based depth recovery systems, however, have not proven practical for many real-world applications. Binocular stereo approaches attempt to mimic the capabilities of the human visual system, and these systems are perhaps the most popular machine vision approaches to the depth recovery problem. However, such approaches are far too computationally burdensome (primarily due to the feature extraction and correspondence steps) to operate in anything close to real time on a conventional processor ([Elf87, Tsa83]). Such approaches also tend to have problems with accuracy, due to feature localization issues and the inherent ambiguity present in the matching process. The motivation for this work is this lack of a general-purpose vision-based range estimation technique capable of operating in real time, while using minimal computational effort.

1.1 Purpose

The purpose of this dissertation is twofold. The first is to present a new technique for recovering depth information from grey-scale imagery which is suitable for real-world tasks. That is, a technique which produces accurate range estimates, while using minimal computational effort. The second aim is to provide a concrete example of how a complex visual task can be simplified considerably by introducing knowledge and control of sensor motion.

1.1.1 THE INTENSITY GRADIENT ANALYSIS ALGORITHM

The Intensity Gradient Analysis (IGA) algorithm is based on the idea that temporal variations in the brightness pattern can be induced in the image, simply by moving the camera. Furthermore, if the motion of the camera is controlled precisely and these temporal variations are measured accurately, is possible to reliably and accurately detect fixed image displacements based solely on these temporal variations. When these fixed displacements are detected, it is possible to recover depth directly, since camera motion is known.

Perhaps the single most appealing aspect of this approach is the minimal computational effort required to recover *accurate* depth estimates. In the IGA paradigm, the depth recovery process is reduced to monitoring the temporal variation in the brightness pattern in a sequence of images acquired using known camera motion. Computationally, this involves a single subtraction operation (to compute the temporal variation in image brightness) and a comparison operation (to see if the fixed displacement has occurred) at those locations in the image where visual depth cues exist. Compare these computational requirements to the feature extraction and correspondence steps inherent in the conventional approaches to the depth recovery problem, and it is not difficult to see the significant reduction in computation effort provided by the IGA approach. This reduction in computational effort (as compared to existing techniques) is significant because there is no apparent tradeoff in accuracy involved in using the IGA approach.

1.1.2 ACTIVE VISION

Recently, researches have begun to realize that by controlling and using knowledge of camera motion, many ill-posed problems in vision become relatively straight-forward. These approaches fall under the loose category of "Active Vision" techniques [AWB88, Baj88, Bal87]. The label "Active Vision," however, seems redundant, as perception is *inherently* an active process. Certainly, in the case of vision, a sensor must at the very least be (actively) directed at an object in order for it to be perceived. And, in many cases, one must (actively) anticipate what one will see in order to correctly interpret the sensory stimulus. Consider Figure 1.1, which shows a classic example of how anticipatory schema come into play. Note that by (passively) looking only at the sensory information, it is not possible to interpret the letter B and the number 13 differently.

This view that perception is an active process is shared by Neisser [Nei76]. He defines perception as consisting of a three-phase cycle in which the perceived information is continually modifying a set of anticipatory schema which, in turn, directs the sensors, which, in turn, perceive the information, thus completing the loop.

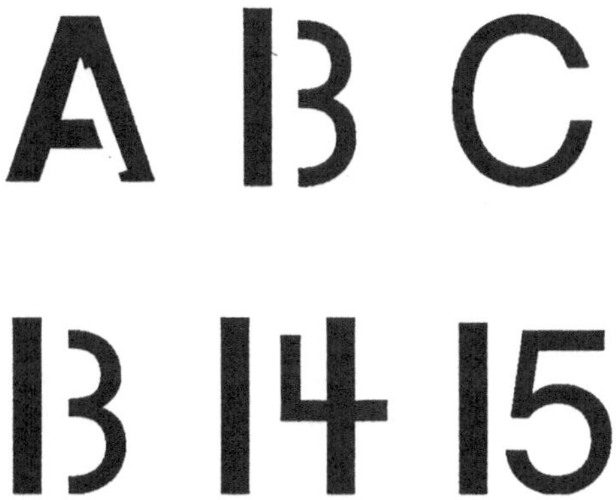

A B C

B 14 15

FIGURE 1.1. A classic example of how one must actively anticipate what one will see in order to correctly perceive sensory stimulus. Note that there is no difference in the sensory stimulus provided by the letter B and the number 13.

> The term *perception* applies properly to the entire (three-phase) cycle and not to any detached part of it [Nei76].

The IGA algorithm uses the additional information available due to knowledge and control of sensor motion and, therefore, perhaps may be categorized as an active vision technique. Semantics aside, however, the approach presented in this thesis is a concrete example of how knowledge about sensor motion can be used to make a difficult problem (depth recovery) quite straight-forward, both intuitively and computationally.

1.2 Philosophy

The intent of the research presented in this dissertation is to develop a robotic vision system. No attempt is made to emulate or model human (or any other animal's) perception in any way. Of course it may be the case that mechanisms similar to that presented in this thesis do exist in nature, but a study of such mechanisms is considered beyond the scope of this research. Certainly, the computer is an excellent tool for studying animal perception and the importance of such research is obvious. However, if the goal is to develop a working robotic system, there is no need to limit one's research by considering only those approaches that are psychologically or physiologically plausible. Of course, there is also no need to explicitly avoid such systems.

Robotic perceptual systems differ from biological systems in many significant ways. The most significant, perhaps, is the fact that robotics systems are very good at processing quantitative information, while biological systems seem more oriented towards qualitative information. Because of this difference, the most efficient and accurate way for computers, or more specifically robots, to process perceptual information may bear little resemblance to the way humans or other animals process this information. Consider, for example, that man's (successful) quest for flight has produced machines that differ from natural flight mechanisms in many significant ways. Ornithopters[1] do exist, but they certainly haven't proven to be the most efficient means of man-made air transportation.

Because the best way for robots to process perceptual information may differ significantly from the way animals do, particular emphasis in the development of IGA has been placed on an "engineering" approach to the depth recovery problem. For example, the first question asked (after "why are existing techniques so slow?") was "Given that we have sensors and computational resources with certain limitations and strengths, how might these sensors and computational resources be best used to recover depth information from an arbitrary scene?" not "This is how humans (or zebras, or emus, or ...) may recover depth, maybe this is how robots should do it."

1.3   The Structure of This Thesis

This thesis can effectively be divided into two parts. Chapters 2 through 5 are more theoretical in nature, describing the depth recovery problem in general the theoretical basis for the Intensity Gradient Analysis (IGA) algorithm, and IGA itself. Chapters 6 through 9 describe the practical aspects of the algorithm: how it is implemented, how one deals with real-world sensors (as opposed to ideal ones), and how the algorithm performs experimentally. Finally, Chapter 10 provides some concluding remarks and discusses the contribution of this work.

---

[1] Ornithopters are machines that propel themselves through the air by flapping their wings.

2

Approaches to the Depth Recovery Problem

In this chapter, the various approaches used to recover depth information are discussed. The first section describes the different sensing modalities that can be used. The second section describes some of the merits and limitations of choosing vision as a primary mode of sensing, and the final section in this chapter is a review of existing vision-based approaches to the depth recovery problem. The primary works in the areas of stereo, optical flow, analysis of spatio-temporal solids, and depth from focus are described, as well as several other approaches. Table 2.1, included at the end of this chapter, provides a qualitative summary of the features and limitations of many of the approaches described in this chapter.

2.1 Sensing Modalities

Two types of sensing modalities exist: active and passive. An active mode of sensing makes some sort of contact with the environment, while a passive mode of sensing perceives the environment without making any contact. Each mode has its benefits and drawbacks and these are summarized in the following discussion.

2.1.1 ACTIVE MODES OF SENSING

In an active mode of sensing, perception is a direct result of an action performed by the sensing device. Without such an action, no information is perceived. Active modes of sensing are desirable because they are (ideally) capable of producing dense depth maps. Since they are not dependent on passive depth cues, it is theoretically possible to perceive the location of an object anywhere in the field of view simply by directing the scanner at it. However, such techniques have some significant drawbacks. Consider the task of determining the location of an arbitrary object using sonar or laser radar. For these techniques, depth is computed by measuring some property of the projected signal after it reflects off the object. The range of these techniques is, therefore, a function of the power of the projected signal and the orientation and reflectivity of the surface of the object. If the projected beam is not powerful enough, no (accurate) signal will be returned. Similarly, if the object is not oriented well, or the surface of the object has poor reflectance properties, little or no signal may be returned to the receiver. Perhaps the most significant limitation of the active approach is the fact that it *is* active in nature. That is, the fact that the agent must make contact with the environment in order to perceive it proves unattractive in many domains. Examples of such domains include those in which multiple agents are operating, where agents probing the environment simultaneously may interfere with each others' signals, or those in which anonymity is important, such as military domains.

Examples of active modes of sensing include tactile sensing, sonar (ultrasound), (laser) radar and structured light techniques ([Bes88] provides an excellent review of active optical range sensors and techniques). Each of these require that some sort of contact be made in order to perceive the environment (whether that contact be physical or electromagnetic). For tactile sensing, perception is the result of making physical contact with the environment. Such sensors are effective only if all objects in the environment are within reach of the sensor. Sonar devices project an acoustic signal and recover spatial information by analyzing the reflections returned from the objects in the environment. Sonar ranging devices produce accurate depth estimates (range accuracy to within 1 percent [ME85]), but they have limited resolution and measuring range (0.9 − 35.0 ft. for Polaroid Sensors [Elf87, ME85]). Laser ranging devices are similar to sonar in that they project a beam into the environment and recover depth information by analyzing the signal reflected by objects in the environment. Laser ranging devices differ from sonar in that the beam projected is produced by a laser, not an acoustic device. These techniques, in general, have excellent accuracy, and as with sonar devices, laser ranging devices are practical only within a limited range ([Bes88]). Structured light techniques use a visual sensor along with a light source that projects a known pattern on the environment. Triangulation is used to determine the locations of points illuminated by the projected pattern. This technique works quite well in controlled environments, such as on a conveyor belt on an assembly line, however, does not generalize well to unstructured environments, where a large depth of field is required.

2.1.2   Passive Modes of Sensing

Passive modes of sensing are desirable because they need make no contact with the environment in order to perceive it: they perceive the environment by sensing stimuli emitted from or reflected off of objects in the environment. Vision, smell and temperature are examples of passive modes of sensing. Both smell and temperature sensors prove to be of little use for the depth recovery application (although it may be argued that the sense of smell may be used under favorable wind conditions to determine the location of the nearest bakery!). Vision is the primary example of a passive mode of sensing useful for recovering structural information from the environment.

As with active modes of sensing, passive modes have their respective advantages. Because of their passive nature, such techniques require the same amount of energy to perceive an object at distance d as they do to perceive an object at distance $100d$. Also, especially in the case of vision, passive techniques often have much higher spatial resolution than active techniques. Certainly, given two techniques that perform equally well, a passive one would be preferable to an active technique. However, passive techniques do have one significant limitation. As with active techniques, passive techniques are capable of recovering information only at locations where perceptual cues exist. For example, if a thermal sensor was used to perceive a room, there would be no way of distinguishing objects from the background when they are at the same temperature as the walls and floor. Only those objects whose temperature is greater than or less than that of the walls and floor will be perceivable. A more detailed discussion of the merits and limitations of one particular passive mode of sensing, namely vision, will be discussed in detail in the following section.

2.2   Vision as a Primary Mode of Sensing

Vision is a particularly appealing mode of sensing for the depth recovery problem for several reasons. A visual sensor perceives the light reflected off of, or emitted by objects in the field of view. Because light is the medium being measured and, ideally, light attenuates very little after traveling over long distances, the effective range of a visual sensor is virtually infinite. Considering that other active techniques have quite limited range, this is a significant advantage. For example, consider the task of constructing an active sensor capable of probing the surface of the moon. Certainly this would be possible, but the power requirements alone would be quite large! On the other hand, a $19.95 dime-store quality camera could be used to perceive the moon with little trouble at all. Of course this is an extreme example, but that fact that the range of a visual sensor is virtually infinite is no trivial matter. Another aspect of vision that is particularly appealing is the fact that visual sensors have excellent spatial resolution (a typical scenario would have a 512 x 512 array perceiving a 60° field of view, or approximately 0.12° per pixel), while ultrasonic sensors have approximately 10° resolution, and laser radar is effectively limited by the time required to scan a scene with high resolution (for example, the Odetics sensor has a pixel dwell time of 32 μsec [Bes88]). Also, a visual sensor needs no moving parts (active sensors must redirect the signal they project to see anything other than a single view).

Vision does have some limitations, however. Being a passive mode of sensing, it is capable of recovering depth information only at those locations where visual depth cues exist. Other limiting factors exist, such as the resolution of the imaging device and the (obvious) fact that vision doesn't work in the dark. Active techniques also have problems with limited resolution, but, since they provide the signal they need in order to perceive, their operation is not limited to domains in which their "perceptual signal" is present (i.e. unlike vision, where ambient light must be present for perception to occur).

Overall, despite these limitations, vision is a very robust mode of sensing with some very desirable properties: it's passive, has virtually infinite range, and has excellent spatial resolution. Because of these attributes, vision seems an ideal sensing modality on which to base a depth recovery system and several techniques do exist for recovering depth information. Unfortunately, however, these visual techniques have not proven practical for real world applications (largely due to their computationally intensive nature). Existing vision-based approaches to the depth recovery problem are described in the following section.

2.3   Literature Survey

Work relevant to vision-based depth recovery falls into two categories: image based techniques, and image-solid techniques. In image-based techniques, information is recovered through the analysis of one or more images. Each image is treated as an individual entity. In image-solid techniques, a large number of images are acquired and a "spatio-temporal solid" is formed with dimensions x, y, and t, where x and y are the dimensions of an image plane and t is time. Examples of image-based techniques include stereo analysis, optical flow analysis, gradient-based analysis, depth from focus and "structure (or shape) from X" approaches. Image-solid techniques are loosely grouped under the heading "the analysis of spatio-temporal solids." A general discussion of each of the above image-based and image-solid techniques is presented, with specific details given for significant contributions to the field.

2.3.1   STEREO

Binocular stereo is perhaps, intuitively, the most appealing technique for recovering depth information from grey-scale imagery. Fundamentally, the computational stereo problem (as perceived by most researchers) is described by the five step process specified by Barnard and Fischler [BF82]:

1. Image acquisition

2. Camera modeling

3. Feature extraction

4. Image matching

5. Depth determination

As *image acquisition* involves the well defined process of receiving information from the sensing device, and *depth determination* is a straightforward task once image matching is done, it is not surprising that the bulk of the research efforts in this area have focused on *camera modeling, feature extraction* and *image matching*.

2.3.2 CAMERA MODELING

The camera model defines the imaging geometry for the system. Since depth values can only be recovered for objects that are contained in the field of view of two (or more) of the sensors, the imaging geometry defines the region in space from which information can be extracted (see Figure 2.1). In addition, the way the camera model is set up is often used to constrain the search for matching points in the images. For example, consider two of the more popular imaging geometries: the binocular stereo model (Figure 2.2), and the axial motion stereo model (Figure 2.3).

The binocular stereo model has been the choice of most researchers [BF82, BK88, Cle87, Gri81, GY88] and certainly is the choice for researchers whose purpose is to study human vision [Gri81, MF81]. In the binocular, or left-right model, the cameras may be aligned so that corresponding scanlines in the two images lie in the same epi-polar plane. This alignment makes it simpler to solve the correspondence problem by limiting the search space to a single dimension.

The axial motion stereo model [OJ84] is often used to exploit known camera motion [IMO84, JBO87]. Animals are known to use a similar technique, called looming, to locate objects (prey, food, etc...) [CH82]. Alvertos, Brzackovic and Gonzalez [ABG89] have shown that by first matching points close to the FOE, fewer ambiguous matches, multiple matches, and false matches occur than in other (conventional) stereo paradigms.

2.3.3 IMAGE MATCHING: THE CORRESPONDENCE PROBLEM

In order to recover depth at a given point in one image (using the conventional approach), the corresponding point must be found in the other image of the image pair. Consider the situation shown in Figure 2.4. The first image (left) shows a set of points in the field of view. We may then be able to infer, given that the sensor moved along its optical axis between acquiring frames, that point A in the first image corresponds to point Q in image two, point D to point S, point E to point T, and so on. However, a problem arises when one tries to find a match for points B and C.

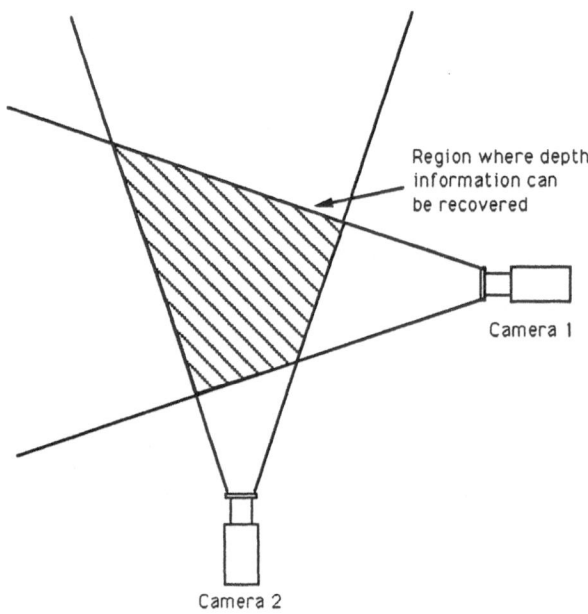

FIGURE 2.1. Range information can only be recovered for objects in the field of view of both sensors (shaded area).

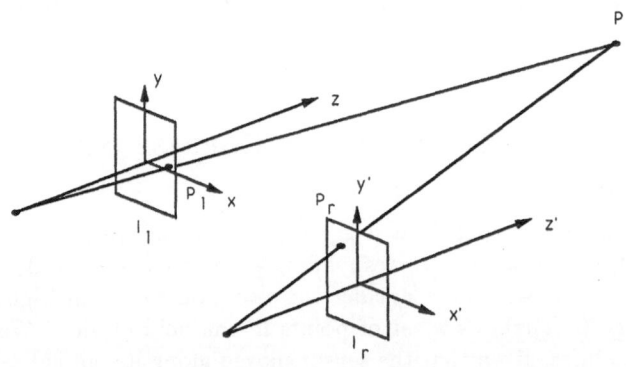

FIGURE 2.2. The binocular stereo camera model.

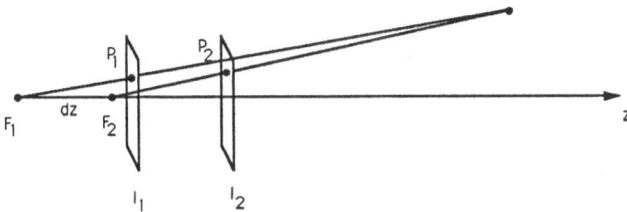

FIGURE 2.3. The axial motion stereo camera model.

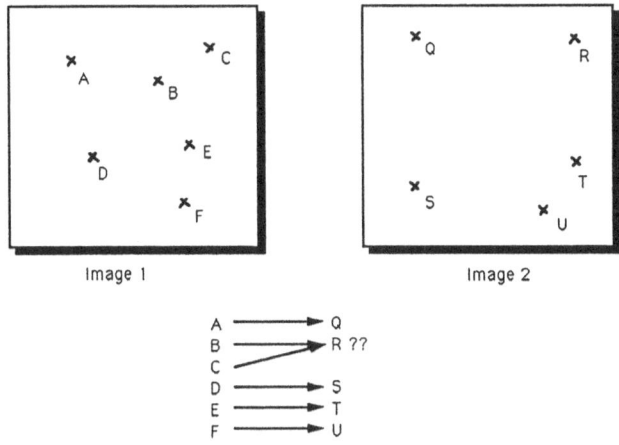

FIGURE 2.4. The correspondence problem: Which point corresponds to R?

Does point B correspond to point R? Does point C correspond to point R? Do they both perhaps correspond to point R? Or, do neither B nor C corresponds to R?

Not surprisingly, a great deal of effort has been focused on simplifying and speeding up the solution of the correspondence problem [BB88, BBM87, Nis84, YKK86, XTA87]. To find the correct correspondences in complicated scenarios, some assumptions about spatial continuity must be made and then some sort of relaxation algorithm can be employed to find the matches [BT80]. Often this technique is applied at several levels of resolution, with the matches found at coarser resolution used to guide the matching process at finer resolutions [Gri81, MP79]. The epi-polar constraint is often used to limit the size of the region searched to find matches. The epi-polar constraint states that corresponding points must lie on epi-polar lines, which are formed by the intersection of the epi-polar plane and the image planes. Small camera movements can also be used to reduce the possible disparities, therefore constraining the area searched to find correspondences [BBM87, XTA87].

2.3.4 FEATURE EXTRACTION

Before correspondences can be determined, a set of points must be extracted from the images. This subset of image points will be used to determine correspondences. Certainly, this subset should be relatively sparse (to facilitate matching), but still capture the "essence" of the image. Often, this subset consists of features such as:

- Zero Crossings of the Laplacian of the Gaussian [Gri81, Nis84]

- Corners [JBO87, Mor81]

- Edges [Cle87, TH87]

Issues involved in selecting such features to look for include operator size, robustness and localization. If speed is a major consideration, the size of the operator should be kept to a minimum because a large operator increases the computational burden on the system. Larger operators, however, tend to be more robust. And, because they have more information available, larger operators tend to give fewer false responses. On the other hand, smaller operators provide better localization, making results more accurate. Smaller operators also tend to produce more feature points, making the correspondence problem more complicated.

2.3.5 COMPUTATIONAL REQUIREMENTS

Combining the problems of feature extraction and correspondence, it is not difficult to see that existing techniques take too much computational effort to be of practical used for many real-world tasks. Consider, for example, Grimson's implementation of the Marr-Poggio algorithm [Gri81]. To determine depth values one must:

- Find the zero-crossings of the Laplacian of the Gaussian of the two images *at four levels of resolution.*

- For each of the four image pairs, find correspondence using the information at lower resolutions to guide the search.

- From the correspondences found at the finest resolution, create a disparity map.

- Using knowledge of the imaging geometry and the disparity map, determine depth values.

Other stereo-based approaches are, generally, all variations on the above theme. Researchers have used different features ([Cle87, JBO87, Mor81, TH87]), projected patterns on the scene to increase the number of matchable features ([Nis84, TK88]), used different baselines (instead of different filter sizes) to guide the matching process ([XTA87]), used different camera geometries ([YKK86]), used correlation measured based on the greyscale surface to determine matches ([WM88]), or, alternatively, a single, moving camera (motion stereo) to obtain images from disparate locations ([JBO87, Nev76, OJ84]). Given the number of computations necessary for feature extraction, the problems with feature localization, and the inherent ambiguity of the matching problem, it is not surprising that these approaches to the depth recovery problem are not practical for most real-world applications.

2.3.6 OPTICAL FLOW-BASED APPROACHES

One particularly appealing approach to the problem of recovering the structure of an arbitrary environment is to compute optical flow. Gibson defines (optical) flow as "... the change (in the image) analyzed as motion perspective"[Gib79]. Another definition, given by Horn and Schunck is "... the distribution of apparent velocities of movement of brightness patterns in an image" [HS81]. Since the brightness patterns in an image arise from objects in the field of view, the apparent velocities of the brightness patterns correspond to the apparent image velocities of objects in the field of view. Therefore, if optical flow is known, the structure of objects in the environment may be deduced.

Most techniques for computing optical flow fall into two categories: gradient-based methods, and token-based methods. Gradient-based techniques attempt to infer motion parameters from image brightness derivatives. Matching techniques locate and attempt to track features over time and, therefore, suffer from the same limitations of the extract-and-match paradigm employed by computational stereo algorithms[KTB87]. Because of this, no further mention will be made of matching techniques. This section discusses gradient-based techniques and some new approaches in which researchers have simplified the problem of computing optical flow by assuming knowledge of the camera motion parameters.

The approach of Horn and Schunck [HS81] is perhaps the best known gradient-based work in this area. To compute the optical flow Horn and Schunck used the motion constraint equation (Equation 2.1), under the assumption that the flow must vary smoothly almost everywhere in the image. The motion constraint equation may be expressed as follows:

$$E_x u + E_y v + E_t = 0 \qquad (2.1)$$

where $E(x, y, t)$ is the image brightness at image location (x, y) at time t, and u and v are the image velocities in the x and y directions, respectively. Although mathematically elegant, this and other gradient-based approaches do not, in general, perform well on real world images. Quantization errors and sensor noise tend to make the computed brightness derivatives both inaccurate and misleading. In general, gradient-based approaches do not perform well with highly textured surfaces, motion boundaries or depth discontinuities [KTB87].

Recently, researchers have simplified the problem of recovering optical flow by taking advantage of knowledge about camera motion [HAH90, MSK89, ST90, VT89]. If one assumes objects in the world are stationary, the flow vectors in the image must originate at the focus of expansion (FOE). Given knowledge of camera motion, we can compute the location of the FOE. Therefore, the problem of recovering optical flow is reduced from determining both the magnitude and orientation of the flow vectors, to simply determining the magnitude of the flow vectors.

Matthies et al. [MSK89] use a simple correlation-based matching algorithm to compute the flow resulting from lateral camera motion. Small camera motions are used to limit the search space, and a four-fold magnification is used to improve precision. Lateral camera motion causes the flow vectors to align with scanlines in the image. This alignment simplifies the search process in the same way the epi-polar constraint is used to simplify the solution of the correspondence problem in stereo-based approaches. Using this simple correlation-based matching approach, by itself, results in limited depth resolution, due to the small baseline between image pairs. However, the Kalman filter is used to assimilate flow estimates over time. The Kalman filter compensates for the limited depth resolution, resulting in accurate depth estimates.

Herman et al. [HAH90] , also use a correlation-based approach to recover the magnitude of the optical flow resulting from lateral camera motion.

Vernon and Tistarelli [VT89] and Sandini and Tistarelli [ST90] use a gradient-based approach to determine the magnitude of the optical flow arising from known camera motion. In this approach, a sequence of images is acquired and convolved with a Laplacian of Gaussian operator. The normal velocities of the brightness patterns in the image at each location on the zero crossing contours are computed using the change in intensity between frames; and the actual velocities are obtained by using knowledge about the camera's motion parameters. As in the approach used by Matthies et al. [MSK89], accurate range estimates are obtained by integrating information from several frames in a sequence. Two types of camera motion are considered in this work: axial motion (motion along the camera's optical axis), and rotation about a fixed point.

2.3.7   Gradient-Based Approaches

Researchers have shown that intensity gradients can be used to infer the structure of an arbitrary environment or recover the camera motion parameters given knowledge about the environment.

Lucas and Kanade [LK85, Luc85] use an intensity gradient-based approach for both the recovery of structural information from an arbitrary environment, and the recovery of camera motion parameters, given a known environment. Their approach involves smoothing the input using a very large mask (up to 65 × 65) and uses the intensity gradients in the resulting smoothed images to register the two frames. A coarse to fine strategy is used, using successively smaller masks and the results at coarser resolution to guide the matching process. A least-squares approach is used to determine correct matches.

This approach suffers from two significant drawbacks. First, due to its multi-scale and iterative nature, it is quite computationally intensive, and, therefore would not be practical for many real-world applications. Secondly, because it bases its decision on the intensity gradients in the region around the point in question, it takes the conventional smoothness constraint one step farther and assumes that intensities must vary (roughly) linearly in the neighborhood of any given point. This, of course may not hold at many interesting areas in the image, especially near object boundaries.

Negahdaripour and Horn [NH86, NH87] use intensity derivatives to recover the motion of an observer relative to a planar surface. Their technique uses spatial and temporal intensity derivatives at a minimum of eight points on a surface and is able to successfully recover observer motion parameters. Although quite elegant mathematically, this technique does not seem to hold much promise for practical implementation for two main reasons. First, it works only for planar surfaces, an assumption that cannot be made for most real-world environments, and, secondly, it assumes that it is working with points from a single planar surface. That is, it assumes that the points selected will lie within the same surface in both frames of the image pair. For real-world applications, this would imply that the segmentation problem would have to be solved before this technique could be applied. If that was the case, there is no sense in using the intensity gradients to recover the camera motion parameters. Such information could be recovered directly from the change in orientation with respect to the segmented objects.

2.3.8 DEPTH FROM FOCUS

Several researchers (including Darrel and Wohn, Grossman, Krotkov and Kories, Pentland, and Subbarao [DW88, Gro87, KK88, Pen87, Sub89]) have proposed methods for recovering depth information by measuring and comparing focus at different lens settings. The basic idea behind depth from focus comes directly from the principles of image formation. In a single lens system, the projection of an object's image on the image plane is governed by the familiar lens equation [Hor86]:

$$\frac{1}{f} = \frac{1}{z} + \frac{1}{\bar{z}}$$ (2.2)

where f is the focal length of the lens, z is the distance from the object to the lens, and $\bar{z}$ is the distance from the lens to the image plane. For a fixed $\bar{z}$, only those points at distance z_0:

$$z_0 = \frac{\bar{z} f}{\bar{z} - f}$$ (2.3)

will be in perfect focus. A point not at distance z_0 from the lens will appear as a circle on the image plane. It is this "blur circle" that is the key to depth recovery using this paradigm.

Again from the lens equation, we can determine the size of the circle projected on the image by a given point. Suppose the point of interest is a distance z_p from the lens. If it were to be in perfect focus, the distance from the image plane to the lens would have to be $\bar{z}_p$, given by:

$$\bar{z}_p = \frac{z_p f}{z_p - f}$$ (2.4)

given $\bar{z}_p$, and the actual distance $\bar{z}$ from the image plane to the lens, we can determine σ, the diameter of the blur circle as:

$$\sigma = \frac{A}{\bar{z}}|\bar{z} - \bar{z}_p|$$ (2.5)

where A is the diameter of the lens[1].

Therefore, the farther an object is from the point of perfect focus (z_0 in our notation), the larger the blur circle for points on that object. And, more importantly, the size of the blur circle is directly dependent on the distance from the object to the observer. Therefore, the fundamental assumption behind depth from focus:

- If one can accurately measure the amount of blur (due to the imaging process), it should be possible to estimate the depth of the object

Of the existing approaches, Grossmann's is perhaps the most intuitively appealing [Gro87]. In this method, a single image is used. The Marr-Hildreth edge operator is applied to find the edges and their orientation. For each edge point, the first derivative of the intensity surface perpendicular to the edge orientation is computed using the original image. From this, a measure W is computed. W is defined as the width of the distribution

---

[1] Note that for each σ there are two possible z_p's

peak of the first derivative. For Grossmann, this is the number of pixels in the peak with values greater than half the maximum derivative. Assuming step edges, a large value for W indicates a large amount of blurring. This process is repeated over several scales to eliminate problems with edges larger (or smaller) than the edge operator. Also, because the estimates for W are inherently quite noisy, these values are averaged over a 16 × 16 window. This technique is quite intuitive in that, for step edges, it is easy to see how W would reflect the nature of the blurring. However, because it assumes only step edges, it is quite easy to see why this technique seems to have little hope for successful implementation on real world images.

Pentland [Pen87] presents two techniques for recovering depth from focus information. His first technique is quite similar to Grossmann's (they appear to have been developed in parallel without knowledge of the other's work), in that a single image is used, edges are detected, and an estimate of the point spread function is computed. This technique is mathematically sound, and is based on modeling the point spread function as a Gaussian distribution, while Grossmann's is ad-hoc. However, it has the same limitations due to the step-edge assumption. Pentland's second approach is based on comparing two or more images of the same scene, acquired using different apertures. The basic idea is that by looking at the frequency spectrum of corresponding patches across different apertures it is possible to uniquely determine the relationship between the point spread functions (blur) in the images. If three or more images are provided, it is shown that the point spread function can be uniquely determined for each image.

There are several ways of implementing Pentland's second approach. The method he selects in his paper he claims is "amenable to an inexpensive, real-time implementation." In this approach, two images are acquired of the same scene. The images are identical, except for their depth of field. The images are convolved with a Laplacian filter, to estimate the high frequency content at each point in the image (the image with the larger depth of field will have more high frequency content due to the sharper edges present). The output of the Laplacian filters are then summed over a small neighborhood and normalized by dividing them by the average grey-level in that neighborhood. A "focal disparity" map is then created by dividing the normalized output of the large aperture (more blurry) image by the normalized output of the small aperture (sharper) image. This focal disparity is analogous to a disparity map created by a stereo algorithm, with the focal disparity monotonically related to depth. Pentland's second approach is much more appealing than the first two because it makes no assumptions about the nature of the edges in the scene, and has potential to be implemented in real or near-real time.

Darrel and Wohn's approach [DW88] uses a sequence of images acquired of a static scene using different focus positions. This approach is similar to Pentland's implementation in that the Laplacian is used to estimate the high frequency content in regions of the image. This high frequency content is compared across the sequence for each region. The central assumption (as with Pentland's approach) is that the output of the Laplacian will be a maximum when the region is in focus. Depth is recovered by recording the focus setting of the lens for the image in which the region has the maximum high frequency content. Depth values are returned only for those regions where a clear maximum high-frequency content exists. This technique introduces two issues not present in the previous approaches. One is the zoom factor – the fact that objects will move in the image due to the varying focus position. The other is the fact that the results are limited to n depth planes, where n is the number of images acquired.

In general, depth from focus techniques suffer from limited depth resolution, and, therefore, have limited usefulness for many real-world applications where depth must be perceived over a large physical range.

2.3.9 STRUCTURE (SHAPE) FROM "X"

Many techniques have been developed to recover the structure or shape of objects in a scene using various cues such as shading, texture and motion. These techniques are mentioned here for the sake of completeness only. For the most part, these approaches are based on regularization and have limited practical applicability due to the ill-posedness of the problem. Texts such as Ballard and Brown [BB82] and Horn [Hor86] provide a good description of these approaches.

2.3.10 ANALYSIS OF SPATIO-TEMPORAL SOLIDS

One way of processing a densely sampled sequence of images is to treat it as a spatio-temporal solid. Bolles, Baker and Marimont [BB88, BBM87] have used this approach to build a three-dimensional description of the environment. By slicing the spatio-temporal solid along the epi-polar lines, epi-polar images are obtained. It is known, by definition, that a feature or object appearing in one image must appear in the intersection of the epi-polar plane on which the object lies and each subsequent image. Therefore, since images are acquired at such a rate so as the temporal continuity between frames approximates the spatial continuity in a single image, these objects or feature points will appear as continuous trajectories in the epi-polar images. Because of this, the correspondence problem is eliminated. Objects are detected simply by detecting their trajectories across the epi-polar images. The locations of objects in the field of view are determined by the orientations of the trajectories in the epi-polar images.

This technique is limited by two factors: first, detecting the trajectories of objects through the spatio-temporal solid becomes non-trivial for any motion other than pure lateral translation, and, secondly, the sheer volume of images required for the technique to work well restricts the speed at which an observer may move through the environment (assuming standard technology image acquisition and processing equipment is used). Also, although it is possible to provide incremental depth estimates based on only a few images using this approach, the accuracy of such estimates is significantly limited by localization problems associated with the edge detection process used to determine object trajectories.

Watson and Ahumada [WA85] applied a complex series of filters to a spatio-temporal solid in an attempt to model human visual motion sensing. Their approach is based on the property of *temporal modularity*, which states that translation at constant velocity changes the static-image Fourier transform into a spectrum that lies in an oblique plane through the origin. Although quite elegant mathematically, this technique has little hope for practical robotic applications due to the extensive computational requirements of the approach.

Following along the lines of the work done by Watson and Ahumada [WA85], Heeger [Hee87] investigated the problem of extracting image flow by computing the motion energy of the spatio-temporal solid by using a 3-D Gabor filter. In this approach, the *motion energy* is computed as the squared sum of the Gabor filter outputs. A gradient descent technique is used to determine the best choice for the flow vector based on the motion energy. This technique seems to work quite well on sequences obtained from highly textured scenes. However, there are two significant limitations of this approach. First of all, the computational requirements are quite large. This technique requires a Gaussian pyramid to be computed for each image in the sequence, each of the resulting spatio-temporal solids is convolved with a 3-D center-surround filter to remove low-frequency components, and the "smoothed" solids are convolved with the Gabor filter to obtain the motion energy. The second limitation of this technique is that it does not lend itself well to incremental update. Because of its computationally intensive nature, it is not practical to compute new results after each image is obtained, one must wait until the entire sequence is available before performing the computations.

Liou and Jain [LJ89] have proposed that motion detection can be accomplished through the analysis of the surfaces formed in spatio-temporal space. They observe that object boundaries will trace out surfaces in a spatio-temporal solid. By computing the surface normal to patches on this surface, velocity information can be recovered. The central contribution of this work is the observation that accurate information can be obtained by working directly in spario-temporal space instead of a transformed frequency domain, which introduces several problems (like quantization error and computational complexity). Although quite appealing conceptually, this technique, as with the previous approaches, is too expensive computationally to be of practical use for many real-world robotic applications.

2.4 Summary

A summary of the features and limitations of the various depth recovery techniques is provided in Table 2.1.

TABLE 2.1. A summary of the features and limitations of various depth recovery techniques.

Features and Limitations of Existing Depth Recovery Techniques

Technique	Features	Limitations

Binocular Stereo	• Mimics the human visual system and is, therefore, intuitively quite appealing,
	
	 • can be used in environments containing moving objects (if cameras are synchronized and frame rate is fast enough),
	
	 • can correctly solve complex visual tasks, such as random-dot stereograms.
		• The inherent ambiguity of the matching process,
		
		 • the extensive computational effort required for:
		
		   − feature extraction,
		
		   − the search for correspondences,
		
		   − the multi-scale nature of the approach,
		
		 • accuracy is dependent on the localization properties of the features used for matching,
		
		 • effective range is limited by the baseline and the resolution of the sensors,
		
		 • produces sparse depth maps,
		
		 • the heuristic nature of the approach.

Technique	Features	Limitations

Token-based Optical Flow		• the extensive computational effort required for:
		
		    – feature extraction,
		
		    – the search for correspondences,
		
		 • the problem is ill-posed (solutions may not converge, or are good only to a scale factor).

Gradient-based Optical Flow	• Can be used to obtain dense depth maps (due to regularization-based nature of the approach).	• Is dependent on accurate estimates of the derivatives of image brightness and, therefore, quite susceptable to quantization errors, • has problems with highly textured surfaces, • flow estimates are instable at depth discontinuities and motion boundaries. • the problem is ill-posed (solutions may not converge, or are good only to a scale factor).

Technique	Features	Limitations
Depth From Focus	• Recovers depth by assuming that depth is proportional to image blur, • Minimal computational requirements, • (some techniques) can be used in environments containing moving objects.	• Limited range over which depth information may be inferred, • dependent on (noisy) estimates of the high-frequency content of the image, • works best in scenes with sharp edges (natural scenes may be a problem).
Image-Solid Techniques	• Interprets a sequence of images as a 3-D spatio-temporal solid, • rigorous mathematical basis, • eliminates the search for correspondences implicit in other vision-based approaches.	• Extensive computational effort required, • limited (or no) trade-off between accuracy and computational effort required, • assumes stationary objects and a moving sensor, or moving objects and a stationary sensor (in most cases).

3

Depth Recovery

This chapter describes the mathematical tools needed to infer depth information from a sequence of two images acquired using a visual sensor undergoing translational motion. The first section of this chapter derives the relationship between the distance to the object, the distance the camera displaced between frames, the location of the object in the image, the orientation of the camera with respect to the axis of translation, and the distance the object displaced in the image between frames. The second and third sections of the chapter discuss the well known axial and lateral camera motion scenarios, respectively. The final section of the chapter summarizes the parameters that must be known in order to recover depth.

3.1   Depth Recovery Using Translational Sensor Motion

Consider the imaging sensor depicted in Figure 3.1. Suppose that the optical system for this camera has a focal length f. The location of a fixed point p in the world cannot be uniquely determined from this single image because there is not enough information present. Suppose, however, that the camera is translated some distance dz, keeping orientation constant with respect to the axis of translation (Figure 3.2). Using basic trigonometry, it is possible to determine the distance z to the point of interest (measured along the axis of translation).

It can be seen that:

$$z \times \tan \alpha = (z - dz) \tan \beta, \tag{3.1}$$

which gives us:

$$z = dz \frac{\tan \beta}{\tan \beta - \tan \alpha}, \tag{3.2}$$

where α and β are as shown in Figure 3.2.

So, in order to recover depth, it is necessary to determine $\tan \alpha$ and $\tan \beta$. From Figure 3.3, we see that:

$$\tan \alpha = \frac{r_1 \sin \theta}{\frac{f}{\cos \phi} - r_1 \cos \theta} \tag{3.3}$$

and, similarly:

$$\tan \beta = \frac{r_2 \sin \theta}{\frac{f}{\cos \phi} - r_2 \cos \theta} \tag{3.4}$$

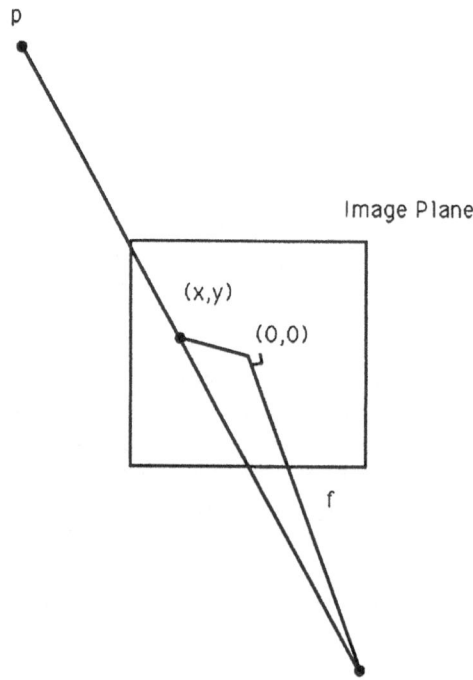

FIGURE 3.1. A simple camera model consisting of the image plane and the focal point of the lens. The optical axis is shown intersecting the image plane at location (0,0). The focal length of this camera is f. Point p in the world is seen at location (x, y) in the image.

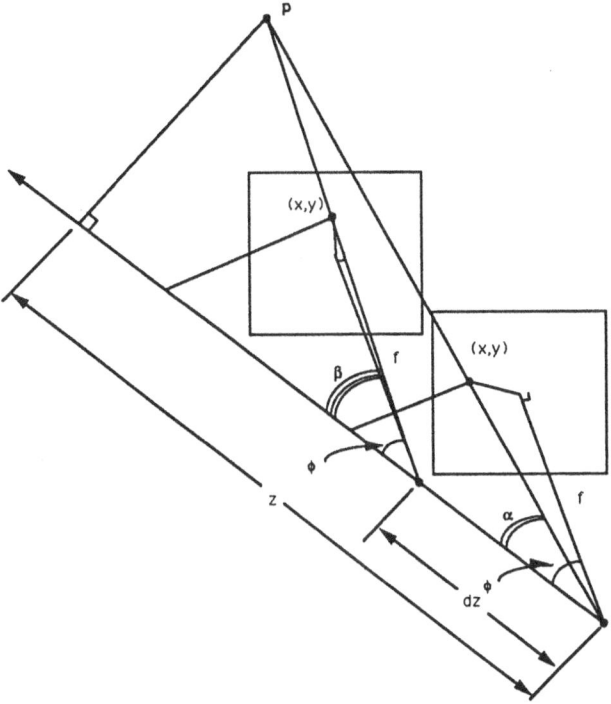

FIGURE 3.2. This figure shows the camera model before and after it is translated a distance dz keeping constant orientation ϕ with respect to the axis of translation. Depth can be computed using dz, α and β.

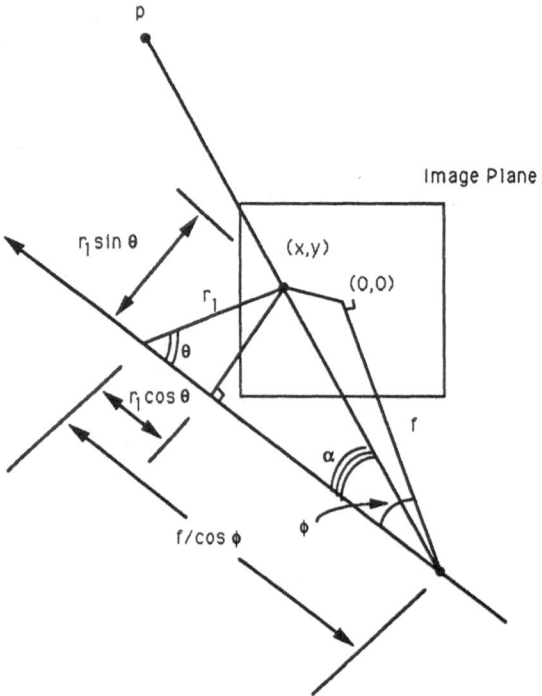

FIGURE 3.3. $\tan \alpha$ can be expressed in terms of the focal length of the lens, the distance r_1 from point (x, y) to the intersection of the axis of translation with the image plane, and θ, the angle between the axis of translation and the line segment connecting point (x, y) and the intersection of the axis of translation with the image plane.

where r_1 and r_2 are the distances from the image of point p to the point where the axis of translation intersects the image plane in the first and second images, respectively, and θ is the angle the image displacement vector makes with the axis of translation.

Substituting Equations 3.3 and 3.4 into Equation 3.2 gives:

$$
\begin{aligned}
z &= dz \frac{\dfrac{r_2 \sin \theta}{\frac{f}{\cos \phi} - r_2 \cos \theta}}{\dfrac{r_2 \sin \theta}{\frac{f}{\cos \phi} - r_2 \cos \theta} - \dfrac{r_1 \sin \theta}{\frac{f}{\cos \phi} - r_1 \cos \theta}} \\
&= dz \frac{r_2 \sin \theta \left(\frac{f}{\cos \phi} - r_1 \cos \theta\right)}{r_2 \sin \theta \left(\frac{f}{\cos \phi} - r_1 \cos \theta\right) - r_1 \sin \theta \left(\frac{f}{\cos \phi} - r_2 \cos \theta\right)}
\end{aligned}
\qquad (3.5)
$$

which reduces to:

$$z = dz\frac{r_2 f - r_1 r_2 \cos\phi \cos\theta}{(r_2 - r_1)f} \tag{3.6}$$

In most cases, it is desirable to represent depth as the distance along the optical axis of the sensor, as opposed to the distance along the axis of displacement. From Figure 3.4, one can see that this distance d can be computed as follows:

$$
\begin{aligned}
d &= (z + z\frac{\tan\alpha}{\tan\theta})\cos\phi \\
&= z(1 + \frac{r_1 \sin\theta}{\frac{f}{\cos\phi} - r_1 \cos\theta} \times \frac{\cos\theta}{\sin\theta})\cos\phi \\
&= z(1 + \frac{r_1 \cos\phi \cos\theta}{f - r_1 \cos\phi \cos\theta})\cos\phi \\
&= z(\frac{f - r_1 \cos\phi \cos\theta + r_1 \cos\phi \cos\theta}{f - r_1 \cos\phi \cos\theta})\cos\phi \\
&= z(\frac{f}{f - r_1 \cos\phi \cos\theta})\cos\phi \\
&= dz\frac{r_2 f - r_1 r_2 \cos\phi \cos\theta}{(r_2 - r_1)f}(\frac{f}{f - r_1 \cos\phi \cos\theta})\cos\phi \\
&= dz\frac{r_2(f - r_1 \cos\phi \cos\theta)}{(r_2 - r_1)f}(\frac{f}{f - r_1 \cos\phi \cos\theta})\cos\phi \\
&= dz\frac{r_2}{r_2 - r_1}\cos\phi \tag{3.7}
\end{aligned}
$$

or,

$$d = dz\frac{r_2}{\delta}\cos\phi \tag{3.8}$$

where δ is disparity ($r_2 - r_1$). Note that, since there is no ambiguity, the distance r_2 from the point p to the intersection of the axis of translation and the image plane in the second image will be referred to as simply r throughout the rest of the discussion.

3.2 Special Case: Axial Camera Motion

When ϕ is 0 (the camera's optical axis is lined up with the axis of translation), equation 3.8 becomes:

$$d = dz\frac{r}{\delta} \tag{3.9}$$

which is the well-known time to collision ratio [Lee76].

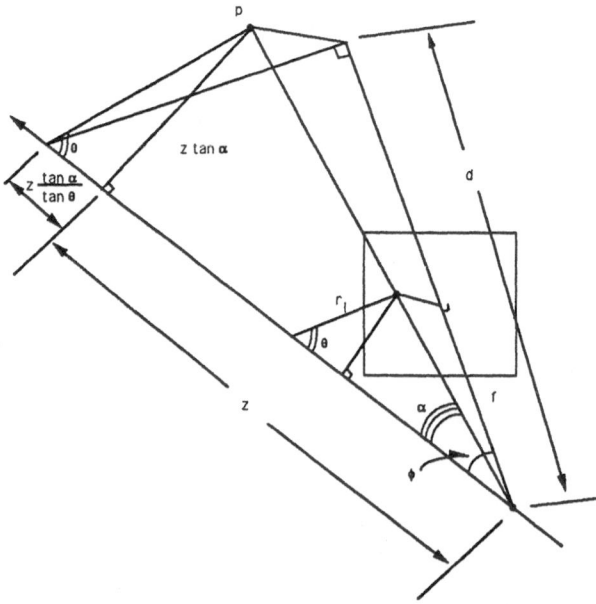

FIGURE 3.4. The distance d to the object, measured along the camera's optical axis can be expressed in terms of z, ϕ, θ and α.

3.3   Special Case: Lateral Camera Motion

When ϕ is 90 (lateral camera motion), r becomes infinite, while $\cos \phi$ goes to 0. However, we know that equation 3.8 must become:

$$d = dz \frac{f}{\delta} \qquad (3.10)$$

for $\phi = 90$ [Hor86]. This implies that $r \cos \phi$ must equal f at $\phi = 90$.

Consider Figure 3.5. Given the location of a point in the image[1] (x, y), the focal length of the lens f, and, assuming (for the sake of simplicity) that the axis of translation intersects the image plane along the line corresponding to $x = 0$ in the image plane, it is possible to determine r as follows:

$$r = \sqrt{y^2 + q^2} \qquad (3.11)$$

where

$$q = f \tan \phi + x \qquad (3.12)$$

which gives us:

$$
\begin{aligned}
r &= \sqrt{y^2 + f^2 \tan^2 \phi + 2fx \tan \phi + x^2} \\
&= \sqrt{y^2 + x^2 + f^2 \frac{\sin^2 \phi}{\cos^2 \phi} + 2fx \frac{\sin \phi}{\cos \phi}} \\
&= \sqrt{\frac{\cos^2 \phi (y^2 + x^2) + f^2 \sin^2 \phi + 2fx \sin \phi \cos \phi}{\cos^2 \phi}} \\
&= \sqrt{\cos^2 \phi (y^2 + x^2) + f^2 \sin^2 \phi + 2fx \sin \phi \cos \phi} \, \frac{1}{\cos \phi} \qquad (3.13)
\end{aligned}
$$

which implies:

$$
\begin{aligned}
r \cos \phi &= \sqrt{\cos^2 \phi (y^2 + x^2) + f^2 \sin^2 \phi + 2fx \sin \phi \cos \phi} \, \frac{1}{\cos \phi} \cos \phi \\
&= \sqrt{\cos^2 \phi (y^2 + x^2) + f^2 \sin^2 \phi + 2fx \sin \phi \cos \phi} \qquad (3.14)
\end{aligned}
$$

which, for $\phi = 90$, equals f, which is the desired result.

3.4   The Parameters Needed for Depth Recovery

In summary, it is possible to recover depth information from two images acquired at disparate locations in the environment using a sensor with fixed

---

[1] Note that image coordinates are set up so that the intersection of the optical axis with the image plane at the origin

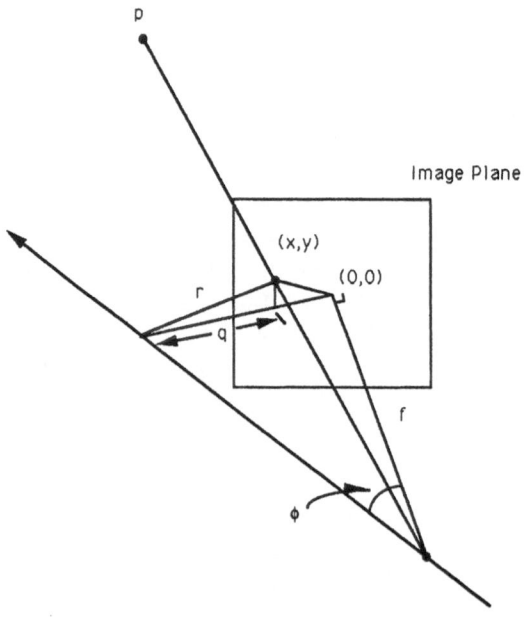

FIGURE 3.5. r can be expressed in terms of the location (x, y) of the object in the image and the distance q, which is the projection of the segment of length r onto the line between the center of the image and the intersection of the image with the axis of translation.

orientation with respect to the axis of motion using the following equation:

$$d = dz \frac{r}{\delta} \cos \phi \qquad (3.15)$$

where d is the distance from the first sensor location to the object (measured along the camera's optical axis), dz is the distance the camera displaced between acquiring the two frames, r is the distance (in the image) from the object point to the FOE, ϕ is the camera's angle of orientation with respect to the axis of translation, and δ is the distance (in the image) that the object displaced between frames.

Note that if camera motion is known, then dz, ϕ and r are also known (recall that r is determined by the location of the FOE, which, in turn, is determined by camera motion). Therefore, the depth recovery problem is reduced to determining the disparity δ.

4

Theoretical Basis for IGA

The previous chapter presented the mathematical mechanisms necessary for recovering depth information from grey-scale imagery. This chapter details the properties of image acquisition and image formation that IGA uses to infer the various parameters used in those mathematical mechanisms.

For input, IGA uses a sequence of images acquired from multiple (known) locations in the environment. This chapter starts with a discussion of two imaging scenarios that can be used to acquire such a sequence: the single-sensor scenario and the multiple-sensor scenario. The relationship between the two is discussed. Although both are useful and relevant to the IGA approach, and each has its advantages, the majority of the discussion in this thesis refers to the single sensor scenario only. The discussion is limited to the single-sensor scenario for the sake of conceptual clarity and it should be noted that assuming a single sensor does not restrict the validity of the derivation.

The second section of this chapter presents two principles of image formation that the IGA algorithm uses to infer depth. The first, called *induced image displacement*, allows one to determine *a priori* the direction in which stationary objects must displace between frames in a sequence. The second, called *temporal coherence*, is needed to infer the behavior of the brightness pattern in the image representing an object in the field of view as the object displaces across the image. This section also explains how temporal coherence and induced image disparity can be used to determine when a specific disparity is seen (and therefore recover depth) without explicitly solving the correspondence problem.

4.1 Acquiring a Sequence of Images

To acquire a sequence of n images from n locations in space, two approaches can be used. The first is to use a single, mobile sensor. To acquire a sequence using this scenario, the sensor is first positioned at location 1 in the environment, image 1 is acquired, then the sensor is moved to location 2 in the environment, image 2 is acquired, and so on, until all n images have been acquired. The second approach is to use multiple sensors fixed in space. In this scenario, image 1 is acquired by sensor 1, image 2 by sensor 2, and so on. Both methods are illustrated in Figure 4.1. Note that the same sequence is obtained in both the single and multiple sensor cases (assuming stationary objects and constant illumination).

In the single-sensor scenario, it is convenient to speak in terms of events

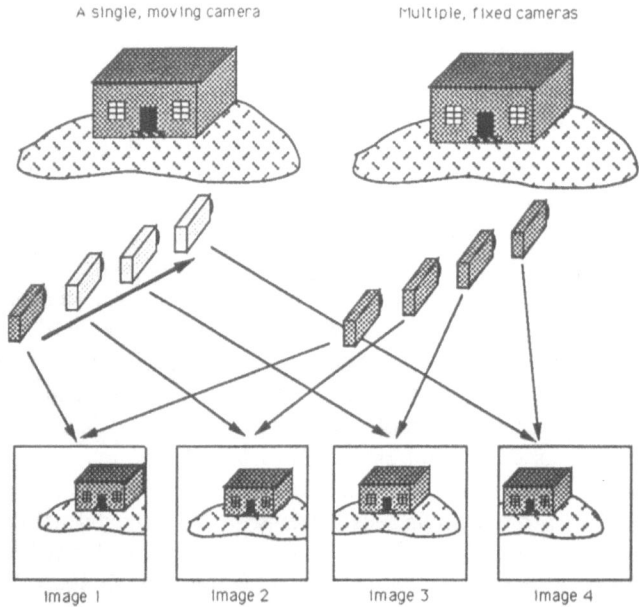

FIGURE 4.1. A sequence acquired using a single, mobile sensor can be equivalent to that acquired using multiple, fixed sensors.

happening over time since the images must be acquired at different time instants. The concept of something happening over time, however, has no meaning for a sequence acquired using the multiple-sensor scenario. The image sequence is the same, though, whether acquired using the single or multiple-sensor scenarios (assuming objects in the environment are stationary). Therefore, "temporal events" in the single-sensor scenario are analogous to "interframe events" in the multiple-sensor scenario.

The majority of the discussion presented in this thesis refers to the single-sensor scenario because it seems more intuitive to define image events as temporal changes perceived on the receptor array of a single sensor. As a matter of convention, the temporal ordering of a sequence acquired using a single sensor will correspond to the spatial ordering of the locations from which the images were acquired. That is, if the sensor moved a distance dz_i between acquiring the first image in the sequence and the ith image in the sequence, and the sensor moved a distance dz_j between acquiring the first image in the sequence and the jth image in the sequence, then $j < i$ implies $dz_j < dz_i$.

4.2   Two Ideas and Their Implications

Essentially, IGA is based on two ideas: induced image displacement, and temporal coherence. The idea of induced image displacement implies that the direction of object motion in the image (the orientation of the displacement vector) can be known at each location in the image. The idea of temporal coherence implies that we know how the brightness pattern perceived in the image must behave if an object displaces between frames. The depth recovery process, or, more specifically, the process of determining disparity, becomes quite straight-forward when the two ideas are combined.

4.2.1   INDUCED IMAGE DISPLACEMENT

A moving camera induces a flow field on the image, causing objects in the field of view to displace in the image. This induced displacement must occur along vectors which originate at the focus of expansion (FOE), and the location of the FOE in the image is determined by the intersection of the image plane with the axis of camera motion (for translational motion). Therefore, if camera motion is known, the direction stationary objects must displace between frames is also known. Finding this direction simply involves determining the location of the FOE with respect to the location of the object in the image.

4.2.2 TEMPORAL COHERENCE

The basic idea behind temporal coherence is that, given the appropriate conditions, if an object displaces in the image, the perceived brightness value at the location the object moves *into* in the current image must equal the perceived brightness at the location the object moved *out of* in the previous image. In other words, the principle of temporal coherence states that the brightness pattern in the image due to an object in the field of view will remain essentially the same for changes in sensor position that are small with respect to the distance to the object. This follows directly from the principles of image formation, and is described in the following discussion.

We know the grey-level I recorded at a given pixel is proportional to the number of photons incident on that region in the image. This relationship can be expressed as:

$$I = k \int \int \rho(x,y) dx dy \tag{4.1}$$

with k being the constant of proportionality and $\rho(x,y)$ being the quantum catch at point (x,y).

Given this relationship, consider the one-dimensional example shown in Figure 4.2(a). Assuming that all pixels have the same spectral sensitivity, we can compute the grey-level recorded at pixel n as follows:

$$I(n) = k \int_{nx_p}^{(n+1)x_p} \rho(x) dx \tag{4.2}$$

where x_p is the width of one pixel.

Suppose that orthographic projection is assumed[1], and that the camera is moved such that the translational component of the induced motion of the object is x_t and that the axial component z_a is very small compared to the distance of any objects in the field of view (making scaling effects insignificant). This is shown in Figure 4.2(b).

We can now compute the grey-level recorded at pixel n from our new camera location:

$$I(n) = k \int_{nx_p - x_t}^{(n+1)x_p - x_t} \rho(x) dx \tag{4.3}$$

$$= k \int_{nx_p}^{(n+1)x_p} \rho(x + x_t) dx \tag{4.4}$$

---

[1]The intent of this discussion is to illustrate the relationship between the object displacement in the image caused by sensor motion and the behavior of the brightness pattern in the image. Orthographic projection is assumed only to simplify the relationship between sensor motion and object displacement. Certainly, perspective projection could be assumed with no loss of generality.

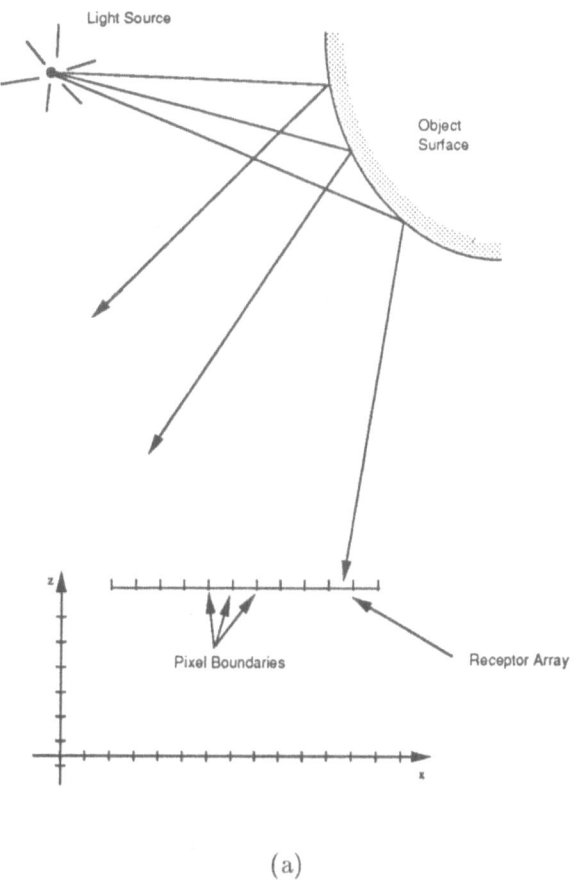

(a)

FIGURE 4.2. A 1-D example illustrating the idea of temporal coherence. (a) A 1-D receptor array before it is moved. (b) The receptor array after it is moved.

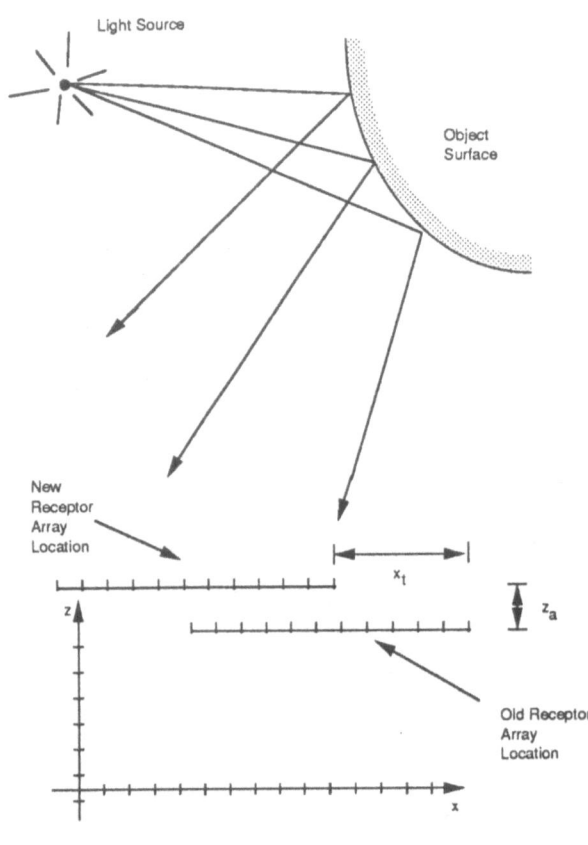

(b)

Let us now look at the special case when $x_t = x_p$.

$$I(n) = k \int_{nx_p}^{(n+1)x_p} \rho(x + x_p)dx \qquad (4.5)$$

$$= k \int_{nx_p - x_p}^{(n+1)x_p - x_p} \rho(x)dx \qquad (4.6)$$

$$= k \int_{(n-1)x_p}^{(n)x_p} \rho(x)dx \qquad (4.7)$$

$$I(n) = I_{prev}(n - 1) \qquad (4.8)$$

Where $I_{prev}(n - 1)$ is the intensity recorded at location $(n - 1)$ before the receptor array was moved. Similarly, if the object displaced the width of k pixels between frames ($x_t = kx_p$):

$$I(n) = I_{prev}(n - k) \qquad (4.9)$$

What this tells us, not surprisingly, is that if an object displaces a distance equal to an integral multiple of the width of one pixel, the intensity perceived at the location the object moved *into* must equal the intensity perceived before the displacement took place at the location the object moved *out of*. This is the principle of temporal coherence.

4.2.3 FINDING DISPARITY WITHOUT SOLVING CORRESPONDENCE

It is not, in general, possible to recover arbitrary disparity without explicitly addressing the correspondence problem. However, by using the two principles outlined above, it turns out that it *is* possible to recover a specific disparity (namely $\delta = kx_p$, where x_p is the width of one pixel and k is an integer) and, thus, recover depth, *without* solving the correspondence problem.

Consider the one-dimensional example shown in Figure 4.3. We know, from the principle of temporal coherence, that the brightness pattern in the image corresponding to the object will appear roughly the same as the object displaces across the image. We also know that, by moving the camera, we can induce an object displacement in the image. Suppose then that the sensor shown in Figure 4.3 is moved to the left. The object must, therefore, displace to the right in the image. Based on the above analysis, when the perceived intensity at pixel location n in image j equals that perceived at pixel location $n - k$ in image 0, it may be possible to assume an image displacement of k pixels has occurred at pixel location n between frames 0 and j. Depth can then be determined using the following equation (See Chapter 3):

$$d = dz\frac{r}{\delta}\cos\phi \qquad (4.10)$$

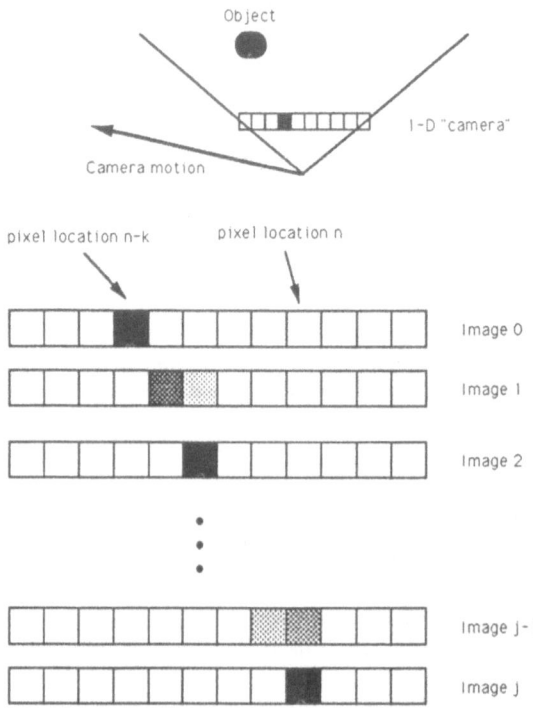

FIGURE 4.3. When the perceived intensity at pixel location n in image j equals that perceived at pixel location $n - k$ on the displacement vector image 0, it may be possible to assume an image displacement of length kx_p has occurred at pixel location n between frames 0 and j.

where dz is the distance the camera displaced between acquiring image 0 and image j, δ equals kx_p, r is the distance from the point in question to the FOE (in image j), and ϕ is the orientation of the camera with respect to the axis of motion.

Note that by simply waiting for this image event to occur, the need to explicitly solve the correspondence problem is eliminated. Correspondence is implicit. No search (and therefore time-consuming computational effort) is required.

Two conditions must be met in order for the above to be true. First, pixel location n in the frame j (and $n - k$ in frame 0) must correspond to a region in the image where a visual depth cue exists. That is, the spatial variation in perceived brightness in the neighborhood of pixel location n must be non-zero in order for this event to imply an image displacement (points in the interior of regions of constant intensity provide no visual depth cues [Gol84]). And, secondly, the grey-level surface between pixel locations n and $n - k$ must be such that, if the intensity I perceived at location $n - k$ in the reference frame is less than the intensity perceived at location n in the reference frame, then the following must hold:

$$I_0(n - k) < I_0(m), \quad n - k < m \leq n. \tag{4.11}$$

Similarly, if $I_0(n - k) > I_0(n)$, the following must hold:

$$I_0(n - k) > I_0(m), \quad n - k < m \leq n. \tag{4.12}$$

This ensures that a k-pixel displacement is uniquely specified by the image event $I_j(n) = I_0(n - k)$. Note that this constraint is similar in spirit to the Nyquist criterion (which defines the minimum sampling rate required to accurately recover a signal with a given maximum frequency component [OS75]). In general, the more frequently (in the spatial domain) the image is sampled (the smaller k is), the more likely it is that the intensity surface will behave as specified above.

Given that the above two conditions are met, specific image displacements can be identified by determining when a perceived brightness value becomes the equal to that perceived a fixed distance closer to the FOE in a previous image.

4.2.4   CORRELATION VS. THE "OPPORTUNISTIC PROCRASTINATION" APPROACH

At first glance, this process may seem like a form of pixel-level correlation. However, there is a significant and important difference. Using a correlation-based approach implies computing a "goodness of fit" measure based on how well the point (or region) in question matches a given mask. The location corresponding to the highest goodness of fit measure is selected as the best match (no matter how poor that match may be). Correlation therefore necessarily implies that a search must be performed to find the best match. The proposed "Opportunistic Procrastination" approach is different from the correlation-based approach in that this approach is purely opportunistic. If the expected image event doesn't occur, *no conclusions are made*. That is, if the match isn't perfect, no "next best" inference is attempted.

Conceptually, perhaps the distinction between the two approaches is best made by considering the following story. A high-level executive process has given two processors the job of determining the magnitude of image displacement. Both are given the perceived brightness value in a reference image at a known image location and both know the location of the FOE. The first processor, P_A (probably a type-A personality), uses a correlation-based approach. First, P_A tells the camera controller to move the sensor some fixed distance. Then, after acquiring a new image, P_A becomes obsessed with his task. He must find the best match, and he must find it now! So, P_A runs frantically up and down the displacement vector (implied by the location of the reference point and the location of the FOE), trying to find the pixel location that best matches his reference value. Once this "best" location is found (no matter how poor a match it is), he computes the distance to the location of the reference value, and runs back to tell the executive process his findings.

Meanwhile, the second processor, P_B, takes a more laid back approach. He takes his reference value (and his beach chair) and moves a fixed distance down the displacement vector (from the reference location) and sits and waits. After making himself comfortable, he tells the camera controller to move the camera some small amount. Then, after acquiring a new image, P_B checks the perceived brightness value at his current location. If this value is equal to his reference value, his task is completed. The image displacement is equal to the distance from his current location to the reference location. Otherwise, he tells the camera controller to move a little more, and repeats the process. He knows the reference value he was given met the two requirements specified above. Therefore, he knows that, if the object corresponding to the reference value is not infinitely far away, the brightness value at his current location must eventually equal his reference value.

Note that, because processor P_B does not try and draw any conclusions if the event he is waiting for doesn't happen, P_B's estimate of image displacement must be as good or better than that obtained using the approach of P_A. And this estimate is obtained with *much less effort* on the part of P_B. Note also that this "opportunistic procrastination" approach combines the principles of least committment and graceful degradation [JH82]. The tradeoff is, of course, that P_B may have to look at more frames in the sequence and P_B must have precise control (or at least knowledge) of the camera motion.

5

Intensity Gradient Analysis

This chapter describes one way of recovering depth information from a sequence of grey-scale images. This technique, called the Intensity Gradient Analysis (IGA) algorithm, is based on the ideas presented in the previous chapter. As one might well infer, IGA uses intensity gradients to determine when a fixed image displacement occurs. Since the IGA algorithm assumes knowledge of camera motion, when a fixed image displacement is perceived, all the parameters needed to determine depth are known.

This chapter begins with a section discussing how a fixed image displacement can be perceived by looking at intensity gradients. The second section of this chapter explains why one would want to do the extra work involved in computing intensity gradients to identify fixed image displacements. The third section of this chapter explains why no further analysis needs to be done in order to apply the methodology presented in Chapter 4 for one-dimensional images to two-dimensional images. In the fourth section of this chapter, the IGA algorithm is presented.

5.1 Isolating Fixed Image Displacements Using Intensity Gradients

To determine if an image displacement equal to an integer multiple of the pixel size has occurred, the previous chapter showed that one must simply determine when the brightness value perceived at some image location equals that perceived at an image location which is a fixed distance closer to the FOE in some previously acquired frame in the sequence. This fixed distance must be an integer multiple of the pixel size. In other words, if an image displacement of magnitude kx_p has occurred at image location n between image i and j (k is an integer and x_p is the width of one pixel), then the following must be true:

$$I_j(n) = I_i(n - k). \tag{5.1}$$

Given this relationship, we can also infer the following: if an image displacement of magnitude kx_p has occurred, then the following must also be true:

$$I_j(n) - I_i(n) = I_i(n - k) - I_i(n), \tag{5.2}$$

where $I_i(n)$ is the brightness value perceived at image location n in image i.

Let the *temporal intensity gradient* at image location (n) between frames i and j be defined as:

$$\frac{\delta I_j(n)}{\delta t} = I_j(n) - I_i(n). \tag{5.3}$$

Similarly, let the *spatial intensity gradient* at image location (n) be defined as:

$$\frac{\delta I_i(n,k)}{\delta n} = I_i(n-k) - I_i(n), \tag{5.4}$$

where $(n-k)$ is the location on the displacement vector $k x_p$ closer to the FOE.

Given these definitions, if an image displacement equal to $k x_p$ occurs, then the *temporal intensity gradient* (change in intensity between frames) at an image location n must equal the *spatial intensity gradient* (change along the displacement vector) at that same location in some previous, or reference, image. That is:

$$\frac{\delta I_j(n)}{\delta t} = \frac{\delta I_i(n,k)}{\delta n} \tag{5.5}$$

or,

$$I_j(n) - I_i(n) = I_i(n-k) - I_i(n), \tag{5.6}$$

for some image j and some previously acquired image i.

5.2   Why Do More Work?

At first glance, computing the spatial and temporal intensity gradients seems like more work than necessary. After all, to determine if a fixed image displacement occurs, one simply needs to compare two values. Why, therefore, would one wish to do the extra computations involved in determining the spatial and temporal intensity gradients? This extra work is needed because of the inherent discrete nature of the approach.

Recall that an image sequence acquired using a moving sensor necessarily consists of a set of images acquired at disparate locations in the environment. Due to the disparate sampling of images, it may be possible that an object displaces less than some fixed distance between the the first (reference) image and, say, frame i, while it displaces more than that fixed distance between the reference image and frame $i+1$. Therefore, it may never be the case that the perceived brightness value at the location in the image corresponding to the object after it has displaced the fixed distance exactly equals the perceived brightness value at that fixed distance closer to the FOE in the reference frame. Thus, by simply monitoring each pixel location and waiting for the perceived brightness value to equal some reference value, *the fixed-distance displacement may never be perceived.*

What this observation implies is that the approach of waiting for a perceived brightness value to equal some reference value may not allow one
to infer depth reliably – some points may be missed. However, by looking
at the spatial and temporal intensity gradients, this failure to perceive a
fixed image displacement may be avoided. Recall condition two from Section 4.2.3. In order to perceive a fixed displacement, the grey-level surface
between pixel locations n and $n - k$ must be such that, if the intensity I
perceived at location $n - k$ in the reference frame is less than the intensity perceived at location n in the reference frame, then the following must
hold:

$$I(n - k) < I(m), \quad n - k < m \leq n. \tag{5.7}$$

Similarly, if $I(n - k) > I(n)$, then the following must hold:

$$I(n - k) > I(m), \quad n - k < m \leq n. \tag{5.8}$$

This condition implies that, for all image displacements less than the fixed
displacement kx_p, the magnitude of the temporal intensity gradient at image location n must be less than the magnitude of the spatial intensity
gradient (in the reference image) at that same image location. Therefore, if
the magnitude of the temporal intensity gradient is greater than or equal to
the magnitude of the spatial intensity gradient, the actual image displacement must be greater than or equal to the fixed displacement the algorithm
is trying to detect.

Following the above line of reasoning, if the magnitude of the temporal
intensity gradient is greater than or equal to the magnitude of the spatial
intensity gradient at some image location a distance r from the FOE, then
the object in the field of view corresponding to this image location must
be located at a distance d from the sensor, where d is defined as follows:

$$d \leq dz_i \cdot \frac{r}{\delta} \cdot \cos \phi, \tag{5.9}$$

where dz_i is the distance the camera moved between acquiring the reference
image and the current image (image i), ϕ is the orientation of the camera's
optical axis with respect to the axis of translation, and δ equals kx_p. Furthermore, if, in the previous image $i - 1$, the temporal intensity gradient at
the same image location was less than the spatial intensity gradient, then
a lower bound can be placed on the estimate of d:

$$dz_{i-1} \cdot \frac{r}{\delta} \cdot \cos \phi < d \leq dz_i \cdot \frac{r}{\delta} \cdot \cos \phi. \tag{5.10}$$

The implications of the above are, although it may not be possible to infer
depth information reliably by waiting for the perceived brightness value
to exactly equal some reference value, it is possible to infer a depth range
in which an object lies by looking at the spatial and temporal intensity
gradients. The size of the depth range is determined by the frequency of
sampling used in acquiring the image sequence, the imaging geometry, and
the location of the object in the field of view.

5.3   Extending to Two Dimensions

Since the previous discussion has been in terms of one-dimensional images, one might think that the technique need be extended to deal with conventional two-dimensional imagery. Such an extension is not necessary, however, since, in this scenario, depth recovery is really a one-dimensional problem. Consider Figure 5.1. Since we know camera motion, we know the FOE. We therefore also know the trajectories the objects must follow. To determine depth at a particular point, we only need know information along its (one-dimensional) displacement vector. Of course, quantization issues do affect the accuracy of the 1-D displacement vectors (due to the discrete nature of the domain), these issues will be discussed in Chapter 6.

An analogy can be made between the displacement vectors from IGA and epi-polar lines in conventional stereo. In both cases, knowledge of imaging geometry is used to reduce depth recovery to a one-dimensional problem.

5.4   The IGA Algorithm

Conceptually, it is convenient to consider the IGA algorithm as consisting of two parts: the initialization stage and the depth recovery loop. In the initialization stage, only the first image from the sequence (the reference image) is processed, and the points where visual depth cues exist are isolated. In the depth recovery loop, fixed image displacements are detected by looking at the temporal intensity gradients between each frame in the sequence and the reference image (note that displacements are measured with respect to the reference image).

5.4.1   THE INITIALIZATION STAGE

In the initialization stage, the IGA algorithm isolates points in the reference image (the first image in the sequence) where depth information may be recovered. Recall that points in regions of uniform intensity provide no visual depth cues; therefore, to isolate those points where depth cues exist, one must simply identify those image locations in regions where the brightness values vary. More specifically, one must isolate those image locations where the brightness values vary along the expected displacement vectors.

Locations where the brightness values vary along the expected displacement vectors can be identified by computing the spatial intensity gradient at each image location (see section 5.1). Recall that the spatial intensity gradient is defined as follows:

$$\frac{\delta I_i(n,k)}{\delta n} = I_i(n-k) - I_i(n), \qquad (5.11)$$

where $I_i(n)$ is the brightness value perceived at image location n in frame

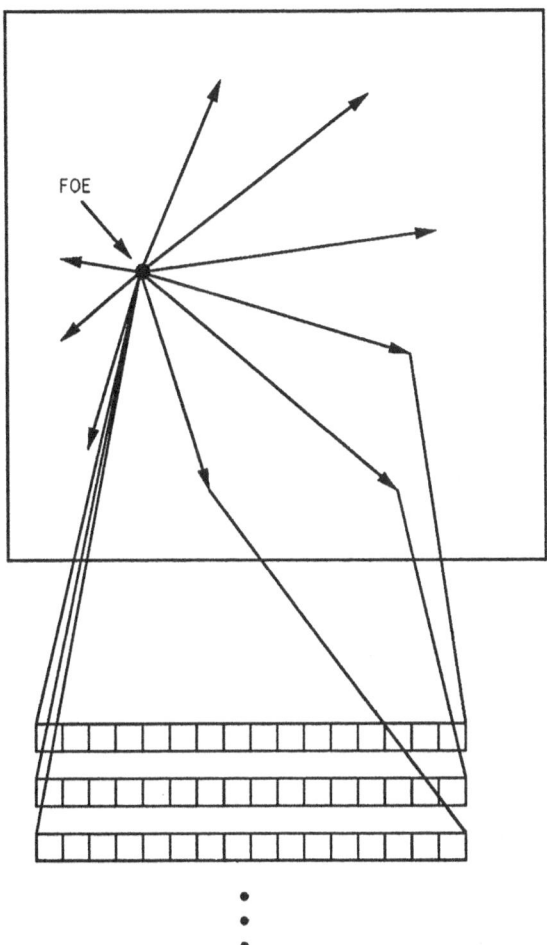

FIGURE 5.1. Applying IGA to two dimensional images involves looking at several one-dimensional problems.

i, and k is the size of the image displacement that is to be detected. If the spatial intensity gradient is non-zero[1], then the point in question must be in a region of varying brightness values and, therefore, may correspond to a visual depth cue.

Using the spatial intensity gradient to isolate points of interest is appealing for two reasons. First, it works. That is, it identifies image locations corresponding to visual depth cues. Secondly, the value of the spatial gradient is needed in the depth recovery loop in order to determine when a fixed image displacement has occurred. Unfortunately, however, selecting the set of *all* points where the spatial intensity gradient is non-zero (or greater than some small threshold) may result in the inclusion of some points that correspond to locations into which regions of constant intensity will displace. Clearly, the temporal behavior of the brightness value at these locations will not accurately indicate a fixed displacement in the image. Therefore, these points must be identified and removed from consideration. Fortunately, these points are easily isolated using a simple test which is described in the following section.

Note that, since camera motion is known, the orientations of the displacement vectors are known at each location in the image. These orientations are collinear with the lines formed by connecting each image location to the location of the FOE in the image. Therefore, for a specific type of sensor motion, the orientations of the displacement vectors are easily determined for each location in the image.

Validation of Visual Depth Cues

The validation stage is necessary to determine if an image location whose spatial intensity gradient is non-zero actually does correspond to an image location where depth can be accurately recovered. Consider a displacement vector whose continuous intensity profile is a step function (see Figure 5.2). It may be the case that the step may not align with a pixel boundary, and, thus, the discrete representation may be a staircase function, as shown in Figure 5.2. In this case, the initialization stage of the algorithm will identify both locations i and $i + 1$ as corresponding to visual depth cues. Unfortunately, however, this situation may lead to an incorrect interpretation of the event when the temporal intensity gradient equals the spatial intensity gradient. Suppose the IGA algorithm is applied to this displacement vector, and an image displacement of one pixel is to be detected ($k = 1$). Note that, at location i on the displacement vector, the brightness pattern need only displace a fraction of the width of a pixel for the temporal intensity

---

[1] In practice, only those points where the spatial intensity gradient is greater than some small threshold are considered for the depth recovery process. This threshold is set to account for the noisy nature of CCD devices. A typical value of this threshold may be 8 greylevels.

Intensity Profile of Displacement Vector (continuous)

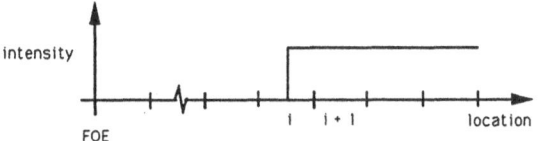

Intensity Profile of Displacement Vector (discrete)

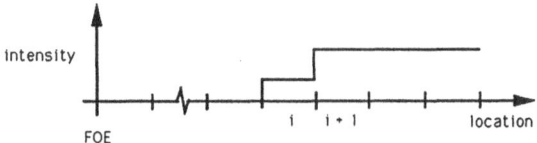

Spatial Intensity Gradient

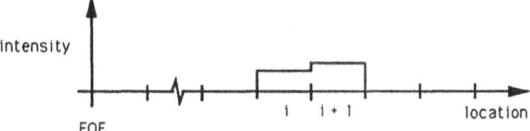

FIGURE 5.2. It may be the case that the boundary of a region of constant brightness may not line up with a pixel boundary. When this misalignment situation
occurs it may be the case that some points corresponding to a non-zero spatial
intensity gradient should *not* be considered for the depth recovery process.

gradient to equal the spatial intensity gradient, while at image location $i+1$
a full pixel displacement must occur if the temporal gradient is to equal
the spatial intensity gradient. This follows from the fact that a region of
constant intensity is displacing into location i, while a region of varying
intensity is displacing into location $i + 1$.

In general, the temporal intensity gradient equaling the spatial gradient will *incorrectly* indicate a fixed image displacement at all points where
boundary of a region of constant brightness does not line up with a pixel
boundary. Fortunately, these points are easily eliminated from consideration by looking at the neighbor of each point considered for depth recovery.
Consider that the spatial gradient is computed by looking at the point in
question and at some image location k pixels *closer* to the FOE. Therefore,
the neighbor, one pixel closer to the FOE, of a point corresponding to the
boundary of a region of constant brightness must not be identified in the
initialization stage as having a non-zero spatial intensity gradient. However, in general, it can be assumed that the pixel boundary will *not* line up

with the boundary of a region of constant brightness (due to simple probability). It may therefore be assumed that the neighbor, one pixel closer
to the FOE, of a point *not* corresponding to the boundary of a region of
constant brightness must have been identified as having a non-zero spatial
intensity gradient. Therefore, only those points whose neighbors, one pixel
closer to the FOE, that also have a non-zero spatial intensity gradient are
considered as points where depth information can potentially be recovered
using this approach.

5.4.2   THE DEPTH RECOVERY LOOP

The depth recovery loop consists of two stages: gradient computation and
depth determination. Images are processed sequentially in temporal order:
the image acquired physically closest to the location where the reference
image was acquired is processed first, then the next closest, and so on.

Temporal Gradient Computation

The first step in the depth recovery loop is to compute the temporal intensity gradient using the current frame and the reference image. Note that the
temporal gradient needs to be computed only at those points corresponding
to visual depth cues; these points were identified in the initialization stage.

If for some image location the magnitude of the temporal gradient is
greater than or equal to the magnitude of the spatial intensity gradient
computed (in the initialization stage) at the same location in the reference image, one can assume an image displacement of k or more pixels has
occurred. This information is passed on to the depth recovery stage. Otherwise, if the magnitude of the temporal gradient is not greater than or equal
to the magnitude of the spatial intensity gradient, then it is concluded that
the object corresponding to that image location must have displaced less
than k pixels between the reference and current frames, and no further
action is taken for that image location in that particular frame.

Depth Determination

In the depth determination step, those points where the magnitude of the
temporal intensity gradient is greater than the magnitude of the spatial
intensity gradient are considered for depth determination. If such a point
has not been considered in this stage of processing before, then a depth
estimate can be computed for that location in the field of view. The high
and low bounds on the depth range corresponding to such a point can then
be determined using equation 5.10, with dz_i corresponding to the camera
displacement used to acquire the current frame and dz_{i-1} corresponding to
the camera displacement used to acquire the previous frame. Otherwise, if
this point has previously been considered at this stage of processing (the
temporal gradient exceeded the spatial gradient for some smaller camera

displacement), no further constraints on the depth range can be made, so no computations are performed.

Note that it may be the case that some previously detected object close to the observer displaces through an image location corresponding to a visual depth cue on some second object. If the second object is much farther from the observer than the first, the displacement of the first (closer) object may cause an incorrect interpretation of the temporal variation in brightness at the reference image location corresponding to the second object. This problem can be avoided by eliminating from consideration all points through which objects that have been previously detected have displaced. The distance each of these objects has displaced can be determined using the following relationship:

$$\delta = dz_i \cdot \frac{r}{d_{min}} \cdot \cos\phi \qquad (5.12)$$

where δ is the distance the object has displaced, and d_{min} is the low bound on the depth range of the object.

The algorithm is given in flow chart form in Figure 5.3

5.4.3 COMPUTATIONAL REQUIREMENTS

As can be inferred from the above analysis, the computational requirements of the IGA algorithm are modest. The initialization stage requires that the spatial intensity gradient be computed at each location in the reference image, and each point at which the spatial gradient is non-zero must be checked to see that its neighbor also has a non-zero spatial intensity gradient. The depth recovery loop requires that the temporal intensity gradient be computed at each image location identified in the initialization stage. If this gradient is greater than the spatial intensity gradient and it is the first time this event has occurred, then the bounds for the depth range can be computed.

Note that, in the very worst case (where *every* pixel in the image corresponds to a visual depth cue – certainly a pathological condition), this is an $O(n)$ algorithm, where n is the number of pixels in an image. A more precise measure of the time complexity of the algorithm is given as follows:

$$T = c_1 \cdot n_p + c_2 \cdot n_{sg} + (F - 1)(c_1 + 2c_2)n_{dq}, \qquad (5.13)$$

where c_1 is the time necessary to perform one subtraction operation, c_2 is the time necessary to perform one comparison operation, n_p is the number of pixels in an image, n_{sg} is the number of points in the reference image whose spatial gradient was non-zero, F is the number of frames in the sequence, and n_{dq} is the number of points corresponding to visual depth cues (the number of points considered in the depth recovery loop). Note that the depth computation can be reduced to table look-up, since a specific image displacement is detected and camera motion is known a priori.

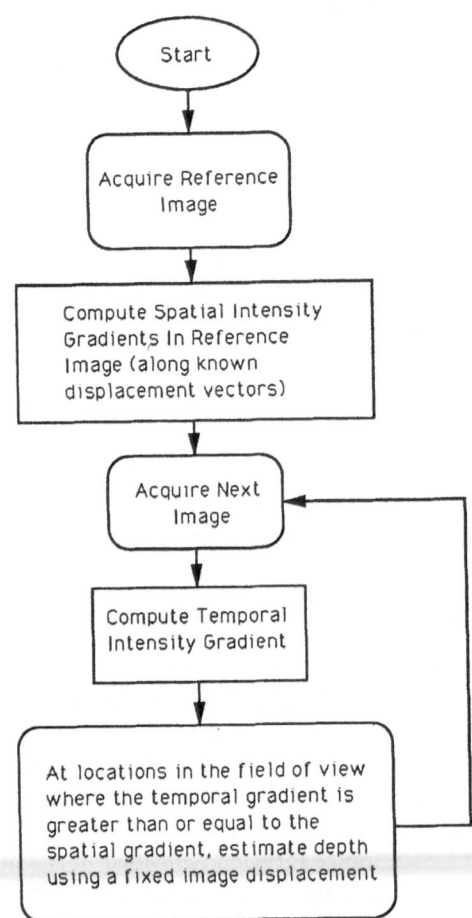

FIGURE 5.3. The IGA algorithm.

6

Implementation Issues

In order to determine if a proposed approach may be useful in real-world applications, it is necessary to understand exactly how the algorithm will perform in the situations it may encounter. A best-case estimate of this performance is given by a theoretical analysis of the approach, assuming ideal conditions. The actual performance of the technique, however, is dependent on how robust the approach is with respect to deviations from ideal conditions. This robustness can be measured experimentally, or predicted by identifying the limitations imposed by real-world implementation. Of course, it is not possible to predict a priori every possible scenario which an approach may be confronted with. Therefore the experimental approach is, at best, incomplete. However, it is possible to predict the performance of the approach based on an understanding of the limitations of the equipment used for implementation and the non-ideal nature of the environment in which the technique is to operate.

Recall that, in the case of the IGA algorithm, depth estimates are obtained by analyzing the temporal variations in the perceived image brightness arising due to the motion of the sensor. Given a perfect sensor, perfect control of sensor motion and an environment consisting of stationary objects only, the IGA algorithm should perform exactly as described in the previous chapter. Unfortunately, however, perfect sensors do not exist and sensor motion may not necessarily be controlled as precisely as one might wish. Therefore, in order to determine the real-world performance potential IGA approach, one must identify the limitations imposed by real-world sensing devices, inexact knowledge of sensor motion parameters, and moving objects. These are each addressed in detail in the following sections. The end of this chapter contains a summary of this analysis.

6.1   Problems with Real-World Sensors

Far too many vision algorithms assume an ideal sensor. Unfortunately, the sensors available today are far from perfect and, although image quality is in general quite good, these images are in fact only discrete approximations to the real thing. This section deals with problems with temporal stability, finite spatial resolution and the geometry of conventional imaging arrays.

FIGURE 6.1. An image showing a typical laboratory scene, acquired using a CCD camera.

6.1.1 TEMPORAL STABILITY OF THE IMAGING SENSOR

Although often assumed otherwise, conventional CCD devices do not have perfect temporal stability. Temporal stability can measured by looking at several frames acquired from the same sensor position and orientation with constant illumination. A sensor with perfect temporal stability would perceive the same brightness value at each image location in each frame in the sequence. Real-world cameras do not have this property. Consider Figure 6.1, acquired using a Sony XC-77 CCD camera. This scene shows a typical laboratory setup. The image quality is, qualitatively, quite good, with good contrast throughout the image. Figure 6.2 shows the same scene, acquired a few seconds later, with no changes in illumination, camera position, or lens parameters. Qualitatively, there is no difference between the frames. However, quantitatively, the difference is quite significant. The difference between the two frames is shown in image form in Figure 6.3. In this image, locations where the perceived brightness varied by one or more greylevel are depicted as black, while those locations where no change occurred are shown in white. Figure 6.4 shows the same image, but with only those locations where the difference in perceived brightness was greater than two greylevels shown in black. A quantitative summary of the results is presented in Table 6.1.

Many factors contribute to the variation in perceived brightness. Beynon and Lamb [BL80] cite three categories of noise sources within the CCD

FIGURE 6.2. The same scene acquired a few seconds later with no change in illumination, camera position or focal parameters. Note that, qualitatively there appears to be no difference between this and the previous frame.

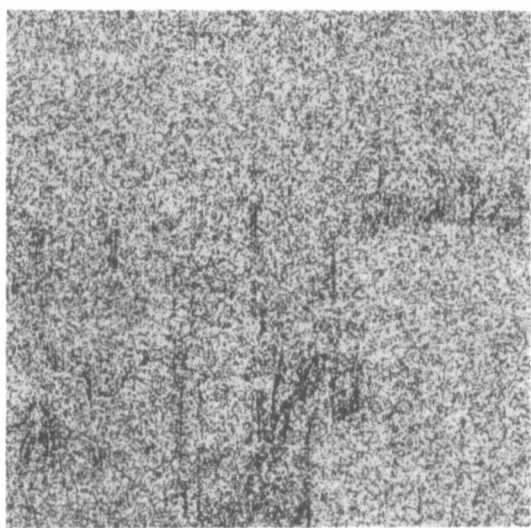

FIGURE 6.3. The absolute difference between the two images. Locations where the perceived brightness changed by one or more greylevel are shown in black.

Absolute Difference	Number of Occurrences
0	84107
1	101773
2	30945
3	5752
4	1445
5	696
6	365
7	221
8	105
9	84
10	35
11	31
12	23
13	3
14	3
15	4
16	5
17	1
18	2

TABLE 6.1. Absolute differences and frequency of occurrence for the laboratory scene (225600 points processed).

FIGURE 6.4. The absolute difference between the two images. Locations where the perceived brightness changed by two or more greylevels are shown in black.

device itself:

- Noise arising from the injection of charge into the device.

- Noise attributed to the fluctuations in the charge transferred from one gate electrode to the next.

- Noise introduced by the charge sensing circuitry.

Other contributing factors may include poor regulation of the power supply, inopportune placement of the cables carrying the video signal, subtle fluctuations in the horizontal sync signal, environmental factors, such as 60Hz light sources, and the inherent data loss resulting from conversion to and re-digitization from the NTSC signal used to transmit the data from the camera to the digitizer.

The above results, although perhaps a little disheartening, are not at all surprising, given the nature of the sensor. CCD imaging devices measure brightness at each image location by measuring the number of electrons liberated by photons incident in a small region of the image plane. The behavior of photons and electrons can, at best, be modeled probabilistically, therefore, it follows that the perceived brightness must behave in a probabilistic manner.

A Statistical Model of Perceived Brightness

Certainly, the temporal stability of the imaging sensor is an important issue in determining how the information is interpreted in an IGA-based depth recovery system. IGA looks directly at the greylevels perceived by the sensor. In order to correctly interpret the variation in greylevels perceived due to objects displacing in the image, one must have a model of the behavior of these greylevels. That is, in order to correctly analyze the output of the algorithm, one must know, for example, if one perceives a change of five greylevels between two frames, what is the probability that one actually *should* have seen such a change? Similarly, if one perceives a change of five greylevels, what is the probability that one should have seen a change of greater than five greylevels? less than five greylevels?

For the purposes of this work, a simple model has been developed for the statistical behavior of the perceived brightness values. In this model, each perceived greylevel is assumed to behave as an independent Gaussian random variable. The Gaussian model was selected because of its simplicity, and, from the Central Limit Theorem, it is known that the distribution of a random variable approaches the normal (Gaussian) distribution as the number of samples gets very large. In other words, for lack of better information, the Gaussian assumption is the best first guess.

It should be reiterated that this model is only an approximation. In reality, the perceived brightness can not in fact be assumed to be an independent random variable. Figure 6.4 illustrates this well. Empirically, it has been found that the variation in the greylevels behaves like a very (very) noisy edge detector. That is, greater temporal variation in greylevels tends to occur at those image locations corresponding to spatial variations in image brightness. From this it can be concluded that the temporal behavior of a particular perceived brightness value is not entirely independent of its neighbors.

Since it is not realistic to model every possible greylevel in every possible situation, an attempt is made only to provide a worst-case estimate of the statistical behavior of a perceived brightness value. Certainly, it may be possible to construct a more accurate model, but this problem is considered beyond the scope of this research.

A worst-case model may be formed by looking at the behavior of the perceived greylevels in the neighborhood of a region where extreme temporal instability is expected. Again, empirically it has been determined that this occurs at regions in the image where large spatial variation in perceived brightness occurs. Therefore, to form a worst case model, the behavior of the greylevels in the neighborhood of a large step discontinuity was studied.

Before the construction of the model is described, a brief review of Gaussian random variables is needed.

The Gaussian probability density function is given in Equation 6.1

([Pee80]):

$$f_x(x) = \frac{1}{\sqrt{2\pi\sigma_x^2}}e^{-(x-\mu_x)/2\sigma_x^2} \tag{6.1}$$

where μ_x, and σ_x are the mean and standard deviation, respectively. Given the probability density function $f_x(x)$, one can, for example, determine the probability of a random variable x taking on a value in some interval (i_1, i_2):

$$P(i_1 < x < i_2) = \int_{i_1}^{i_2} f_x(x)dx \tag{6.2}$$

This fact will be used shortly.

Of course, to use the Gaussian model, one must know the parameters μ_x and σ_x. These can be determined experimentally using the following:

$$\mu_x = E[x] \tag{6.3}$$

and,

$$\sigma_x = E[(x - \mu_x)^2] \tag{6.4}$$

where $E[\cdot]$ is the expectation operator.

To determine the values of μ_x and σ_x, one thousand images were acquired of a scene similar to that depicted in Figure 6.5. The grey-values in a 10×10 neighborhood about the black to white transition on the target object were analyzed. Figures 6.6 and 6.7 show the μ_x and σ_x values, respectively, for the 10×10 region.

As can be seen from Figure 6.7, the standard deviation (which determines the shape of the distribution function) ranges from 0.65 at subimage location (9,4) (top left is (0,0)), to 5.58 at subimage location (6,1). Assuming that $\sigma_x = 5.58$ is a worst-case estimate, what are the implications of this with respect to the calculations performed by the IGA algorithm? Recall from Chapter 5 that the IGA algorithm attempts to determine when the temporal variation in intensity between frames in a sequence is equal to or greater than the spatial variation in intensity at that same location in the reference image (the first frame in the sequence). Mathematically, this can be expressed as follows:

$$|I_t(x) - I_0(x)| \geq |I_0(x - k) - I_0(x)| \;\wedge$$
$$sgn(I_t(x) - I_0(x)) = sgn(I_0(x - k) - I_0(x)) \tag{6.5}$$

where $I_0(x)$ is the greylevel perceived at location x on the displacement vector in the reference frame, $I_t(x)$ is the greylevel perceived at location x on the displacement vector in frame t, and $\wedge$ is the and operator. For the sake of argument, assume that $I_0(x) < I_0(x - k)$. Let λ be defined as follows:

$$\lambda = I_t(x) - I_0(x - k) \tag{6.6}$$

FIGURE 6.5. The target used to compute the worst-case behavior of the perceived brightness. A 10 × 10 region centered on the black to white transition on the target was analyzed.

199.0	199.0	197.9	197.1	194.6	182.8	123.9	50.3	40.1	43.3
200.2	199.1	197.7	195.6	193.3	180.8	120.0	49.2	40.3	43.4
199.2	198.4	197.0	194.7	193.1	179.5	116.0	47.9	40.3	43.6
200.5	199.4	197.6	195.0	193.2	178.1	112.6	47.1	40.5	43.5
199.5	198.5	196.5	194.8	192.7	177.4	110.1	46.6	40.4	43.5
198.9	197.1	195.7	194.5	192.9	176.5	106.9	45.6	40.5	43.3
199.9	199.3	197.9	195.5	194.1	175.6	103.8	45.3	40.9	43.1
200.1	198.7	197.9	196.7	194.3	173.2	98.3	44.0	41.1	43.4
199.2	197.2	195.0	193.7	192.9	172.6	96.3	43.4	41.2	43.4
198.9	197.7	196.2	195.2	193.0	169.9	91.1	42.8	41.9	43.8

FIGURE 6.6. Mean values (μ_x) for the 10 × 10 window in the target image.

1.01	1.02	1.04	1.02	1.02	1.83	5.56	2.21	0.75	0.68
1.08	1.07	1.06	0.99	0.98	2.06	5.58	2.15	0.82	0.71
0.98	1.03	1.10	1.06	1.04	2.23	5.41	1.94	0.82	0.66
1.03	1.06	1.10	1.07	1.06	2.34	5.36	1.85	0.80	0.68
1.07	1.06	1.05	1.02	1.00	2.52	5.36	1.76	0.82	0.65
1.06	1.02	1.04	1.03	1.02	2.63	5.28	1.67	0.81	0.67
1.07	1.03	1.04	1.03	1.03	2.99	5.39	1.50	0.78	0.66
1.03	0.99	1.01	1.06	1.04	3.20	5.13	1.38	0.78	0.68
1.04	1.06	1.03	1.02	1.02	3.41	5.21	1.30	0.77	0.70
1.04	1.04	1.06	1.05	1.06	3.67	4.98	1.10	0.70	0.68

FIGURE 6.7. Standard deviations (σ_x) for the 10 × 10 window in the target image.

In this scenario IGA is monitoring image location x, waiting for λ to become greater than or equal to zero.

Assuming that $I_0(x)$ and $I_t(x)$ are Gaussian random variables with standard deviation $\sigma_I = 5.58$, one can conclude that λ is also a Gaussian random variable with $\sigma_\lambda = 7.89$ (recall that the variance of a sum of n random variables with variance σ^2 is $n\sigma^2$).

Given σ_λ and recalling Equation 6.2, one can determine the probability that, given an observed lambda value, λ_i, the actual value of lambda was greater than or equal to zero. Consider an observed λ_i. Suppose the expected value of λ is zero ($\mu_\lambda = 0.0$). The probability density that the observed value is λ_i is simply $f_\lambda(\lambda_i)$. Since we are interested in determining when the actual value of λ is greater than or equal to zero, we must also consider the case where the the expected value of λ is greater than zero ($\mu_\lambda > 0.0$). In this case, the probability density that the observed value is λ_i is $f_\lambda(\lambda_i - \mu_\lambda)$. Following this line of reasoning, one obtains $P(\lambda \geq 0|\lambda_i)$, the probability that the actual value of lambda was greater than or equal to zero, given an observed value of λ_i as follows:

$$P(\lambda \geq 0|\lambda_i) = \int_0^\infty f_\lambda(\lambda_i - x)dx$$

$$= \int_{-\infty}^{\lambda_i} f_\lambda(x)dx \qquad (6.7)$$

which has no known closed-form solution, but is tabularized in many references [Pee80].

Using Equation 6.7 one can construct a table showing the probability that λ is actually greater than or equal to zero, given that the observed value was λ_i (see Table 6.2). A similar table can be constructed for the case where $I_0(x) > I_0(x - k)$, this would be similar to Table 6.2, but the

λ_i	$P(\lambda \geq 0.0)$	λ_i	$P(\lambda \geq 0.0)$
-10	0.104	0	0.500
-9	0.127	1	0.591
-8	0.156	2	0.599
-7	0.187	3	0.648
-6	0.224	4	0.695
-5	0.264	5	0.736
-4	0.305	6	0.776
-3	0.352	7	0.813
-2	0.401	8	0.844
-1	0.481	9	0.873
		10	0.896

TABLE 6.2. The probability that λ was actually greater than or equal to zero, given the observed value λ_i.

probabilities associated with positive λ_i values would be swapped with the probabilities associated with the negative λ_i values.

From Table 6.2, one can conclude that, in the worst case, the temporal stability of the CCD camera is not good enough to accurately infer information based on the temporal variation in perceived brightness. Consider a perceived temporal variation of 15 ($I_t(x) - I_0(x) = 15$), and suppose the spatial variation in the reference image was 20 ($I_0(x - k) - I_0(x) = 20$). In this case, the observed value of lambda is -5 ($\lambda_i = -5$). From Table 6.2 we can conclude that there is better than a one-in-four chance that the actual value of lambda was really zero or greater. Similarly, if $\lambda_i = -10$, then there is better than a one-in-ten chance that lambda was really zero or greater. Given that many visual depth cues are composed of spatial variations in intensity of only a few greylevels, the temporal stability of the sensor used is clearly unacceptable[1].

Smoothing the Input

One way to accommodate for the poor temporal stability of the image sensor is to filter the image using a smoothing algorithm. Such an algorithm modifies each pixel value by setting it equal to some function of the pixel values in the neighborhood of the pixel in question. Examples of smoothing techniques include Gaussian filtering, median filtering, and

---

[1] Note that these results are indicative of the performance of CCD devices in general, and shouldn't be construed to imply that the XC-77 is a poor-quality imaging device. On the contrary, the Sony camera used for these experiments had, in fact, better temporal stability than other cameras tested.

Smoothing Window Size	σ_{λ_s}	
	predicted	measured
2×2	3.95	3.05
3×3	2.63	-
4×4	1.97	1.93
5×5	1.57	-
6×6	1.32	-
7×7	1.12	-
8×8	0.98	0.971
9×9	0.88	-
10×10	0.79	-

TABLE 6.3. Smoothing window sizes and the predicted values for σ_{λ_s}.

simple averaging. An averaging technique has been selected for use in this work. This technique computes the average of the perceived brightness values in a neighborhood about each point. More complex (and perhaps more appropriate) algorithms may exist, but the averaging approach has proven sufficient. The issue of determining the optimal filter for use with this approach is considered beyond the scope of this work.

One would expect the temporal stability of the smoothed input to be significantly better than that of the raw input. This follows from the fact that the standard deviation of the average of N uniformly distributed values is simply the standard deviation of one value divided by the square root of N. Let λ_s be defined as follows:

$$\lambda_s = \mathcal{I}_t(x) - \mathcal{I}_0(x - k) \tag{6.8}$$

where $\mathcal{I}_t$ is the smoothed input at time t. Again assuming that $\mathcal{I}_0(x) < \mathcal{I}_0(x - 1)$, IGA will monitor image location x, waiting for λ_s to become greater than or equal to zero.

Recalling that the original input had variance $\sigma_I = 5.58$, one can predict the variance of λ_s for various smoothing window sizes. This value was also determined experimentally for 2×2, 4×4, and 8×8 filter sizes (using the same 1000 images processed to compute σ_I). These values are shown in Table 6.3.

Given the value for σ_{λ_s}, one can, using Equation 6.7, construct a table showing the probability that λ_s is greater than or equal to zero for a given observed lambda λ_i and a specific sized smoothing window. These probabilities are given in Table 6.4 for 2×2, 4×4 and 8×8 filter sizes. As expected, performance improves as the filter size increases.

λ_i	$P(\lambda \geq 0.0)$ window size			
	no smoothing	2×2	4×4	8×8
-10	0.104	0.006	0.0	0.0
-9	0.127	0.011	0.0	0.0
-8	0.156	0.022	0.0	0.0
-7	0.187	0.038	0.0	0.0
-6	0.224	0.064	0.0	0.0
-5	0.264	0.104	0.005	0.0
-4	0.305	0.156	0.021	0.0
-3	0.352	0.224	0.064	0.0
-2	0.401	0.305	0.154	0.021
-1	0.481	0.401	0.305	0.151
0	0.500	0.500	0.500	0.500
1	0.591	0.599	0.695	0.849
2	0.599	0.695	0.846	0.979
3	0.648	0.776	0.936	0.999
4	0.695	0.844	0.979	0.999
5	0.736	0.896	0.995	0.999
6	0.776	0.936	0.999	0.999
7	0.813	0.962	0.999	0.999
8	0.844	0.978	0.999	0.999
9	0.873	0.989	0.999	0.999
10	0.896	0.994	0.999	0.999

TABLE 6.4. The probability that λ_s was actually greater than or equal to zero, given the observed value λ_i for several different smoothing window sizes.

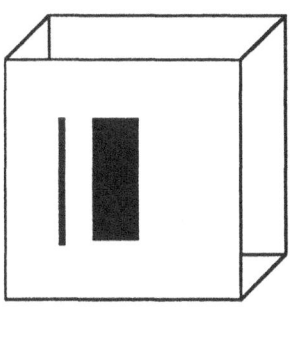

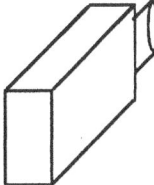

FIGURE 6.8. An example imaging situation where the spatial resolution of the sensor proves insufficient.

6.1.2 FINITE SPATIAL RESOLUTION

An image acquired using a CCD sensor is a discrete approximation of the view actually seen by the camera. Although it is conceptually convenient to assume that each pixel location corresponds to a point in the field of view, this is, in truth, not a valid assumption. Pixels have finite size. Therefore, the brightness recorded at each pixel location corresponds to the brightness in a region rather than at a point. Because of this, the behavior of the greylevels in the image may not accurately reflect the behavior or nature of objects in the field of view. Consider the following situation: there is a stationary object in the field of view the surface of which is marked with two vertical black lines, one thick one, and one thin one (see Figure 6.8). The geometry of the imaging situation is such that the thick line is much wider than the width of a pixel and the thin line is thinner than the width of one pixel (see Figure 6.9). The discrete image is shown in Figure 6.10. Note that the image locations corresponding to the thin line and the edge of the thick line are dark grey and not black, like the actual pattern on the object.

Not only do the limited resolution images inaccurately reflect the nature of the surfaces of objects, but, more importantly (in terms of this work), they also may inaccurately reflect the behavior of objects as they displace in the image. A fundamental assumption for this and other motion-related image analysis is that the apparent velocities of the brightness patterns in the image correspond to the apparent image velocities of objects in the

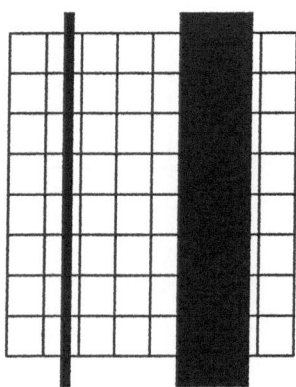

FIGURE 6.9. The discrete image grid superimposed over the actual scen

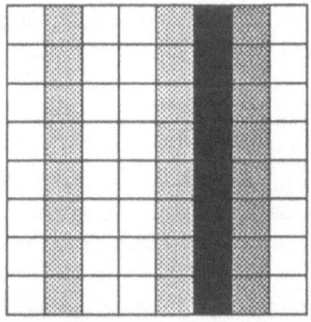

FIGURE 6.10. The discrete image of the scen

field of view. Unfortunately, due to limited resolution, this is not always the case. Consider again the scenario depicted in Figures 6.8-6.10. Since the projected width of the thin line is less than the width of one pixel, it may be possible that a perceived image displacement of one pixel may correspond to an actual physical displacement of less than one pixel. This is illustrated in Figure 6.11.

It should be noted that errors caused by the finite resolution of the imaging sensor will always result in "safe" mistakes. That is, a perceived image displacement will always be greater than or equal to the actual image displacement. Therefore, since perceived depth is inversely proportional to perceived image displacement, the errors in the depth estimate due to limited spatial resolution will always be on the conservative side, indicating that objects are closer than they actually are.

Figures 6.12 - 6.14 show the results of an experiment illustrating this phenomenon. Figure 6.12 shows a 480 × 480 image of a target that is located 170 cm from the observer. The target consists of several vertical bars of varying widths. Figure 6.13 shows the same scene, but at one quarter the resolution (120 × 120). Note that the narrower bars towards the right of the target lose their definition in the lower resolution image. Figure 6.14 shows the output of the IGA algorithm when applied to a sequence acquired at the lower (120 × 120) resolution. Note that the region of the output image on the right side of the target (where the poor spatial resolution is evident in Figure 6.13) is darker than the rest of the target area. This indicates that the algorithm perceived this region as being closer than the rest of the target.

6.1.3 GEOMETRY OF CONVENTIONAL IMAGING ARRAYS

Since conventional imaging arrays are arranged in a row-column format, they do not lend themselves well to the analysis of arbitrary displacement vectors (other than those with orientations that are multiples of $\frac{\pi}{2}$). Suppose, for example, axial camera motion is used to acquire a sequence of images, and the center of the image is at location (i, j). In this case, the displacement vectors in row i and column j line up properly with the imaging array (see Figure 6.15). However, for pixels that lie off of row i and column j, say at image location $(i - n, j - m)$, the discrete representation of the continuous displacement vector may be a poor approximation (see Figure 6.16).

Suppose an image displacement of one pixel is to be detected. In this particular example, the location on the discrete displacement vector used to compute the spatial intensity gradient is image location $(i-n+1, j-m)$. Unfortunately, as can be seen in Figure 6.17 there are two main problems with using pixel $(i-n+1, j-m)$ to compute the spatial intensity gradient:

1. Pixel $(i-n+1, j-m)$ is different from the location on the displacement

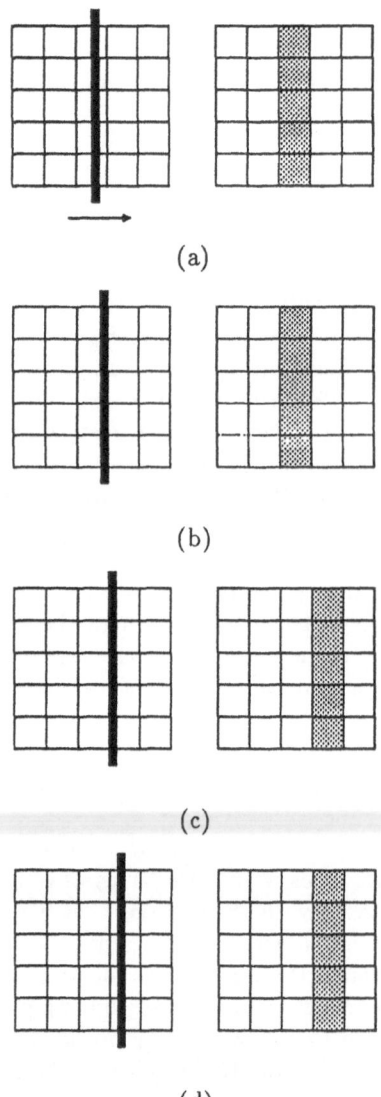

FIGURE 6.11. An example where a perceived image displacement of one pixel corresponds to an actual physical image displacement of less than one pixel. (a) actual displacement = 0, perceived displacement = 0. (b) actual displacement = 0.25, perceived displacement = 0. (c) actual displacement = 0.5, perceived displacement = 1.0. (d) actual displacement = 0.75, perceived displacement = 1.0.

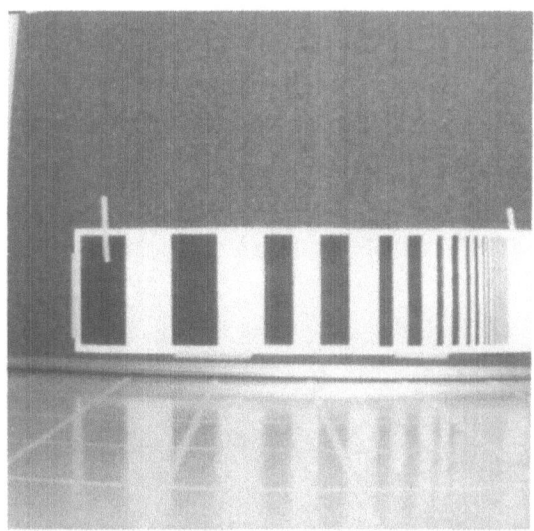

FIGURE 6.12. The target used to test the effect of spatial resolution (480 × 480 resolution). The target is 170cm from the observer.

FIGURE 6.13. The target used to test the effect of spatial resolution (120 × 120 resolution).

FIGURE 6.14. The output of the IGA algorithm applied to the spatial resolution sequence (120 × 120 resolution. Depth is encoded as brightness. The closer the object, the darker it appears. White indicates no information recovered). The algorithm correctly locates the target at all points except those where the spatial frequency of features is greater than the resolution of the imaging device.

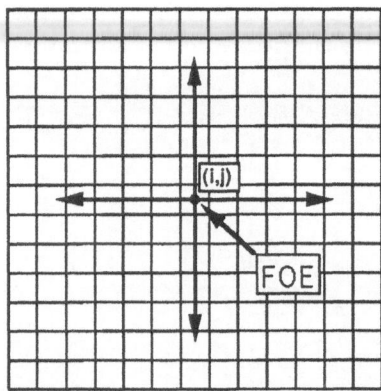

FIGURE 6.15. Using a conventional imaging array, only those displacement vectors that align with the row or column on which the FOE lies will be properly represented in discrete form.

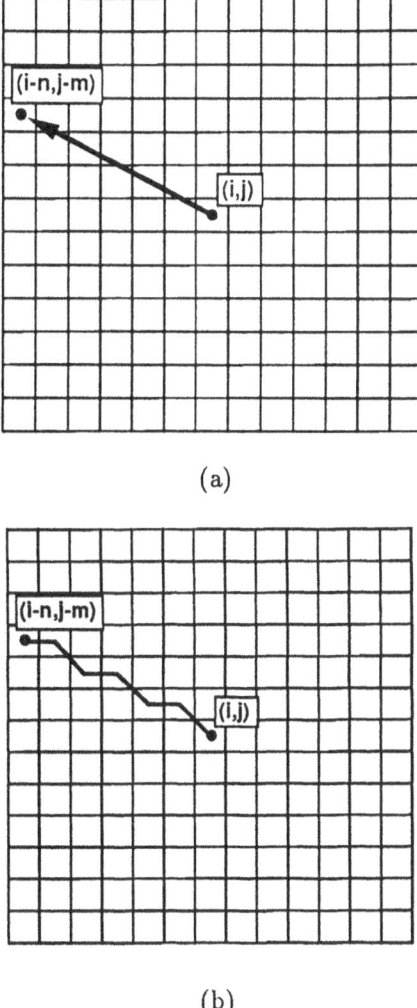

(a)

(b)

FIGURE 6.16. The discrete representation of a continuous displacement vector may be a poor approximation ((b) and (a) above, respectively).

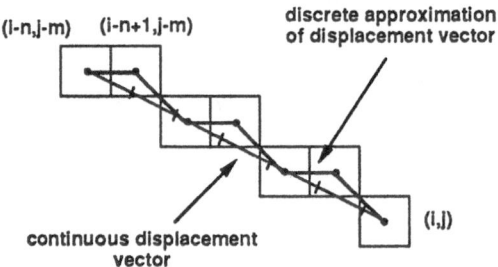

FIGURE 6.17. Problems associated with the discrete approximation to a continuous displacement vector (see text for a complete description).

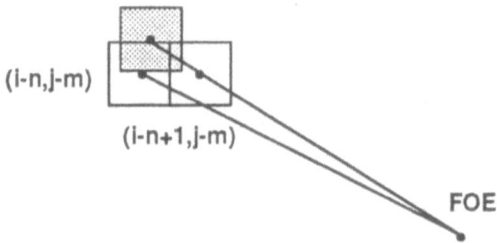

FIGURE 6.18. The shaded area shows the portion of the scene perceived at pixel location $(i - n + 1, j - m)$ in the reference image after it has displaced. Note that this region only partially overlaps pixel location $(i - n, j - m)$, making the spatial gradient computed at pixel location $(i - n, j - m)$ an inaccurate estimate of the temporal variation needed to indicate a single pixel displacement.

vector 1 pixel width closer to the FOE from pixel location $(i - n, j - m)$.

2. The orientation of pixel $(i - n + 1, j - m)$ with respect to pixel $(i - n, j - m)$ is such that the region of the scene perceived at pixel location $(i - n + 1, j - m)$ will displace only partially into the region of the scene perceived at pixel location $(i - n, j - m)$, therefore making the spatial gradient computed at pixel location $(i - n, j - m)$ an inaccurate estimate of the temporal variation needed to indicate a single pixel displacement (this is shown in Figure 6.18).

One way of limiting the deleterious effects of discretization is to consider larger image displacements. That is, look for image displacements of more

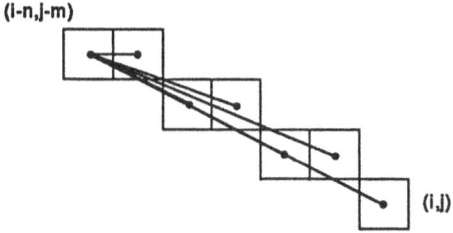

FIGURE 6.19. The discrete approximation of the displacement vector improves in accuracy as larger image displacements are considered.

than one pixel. An improvement in performance is expected because the maximum possible error in orientation of the discrete displacement vector is defined by the ratio of $\sqrt{2}$ times half the width of one pixel (the maximum possible distance a point can be from the center of a pixel and still be in the pixel boundary) to the magnitude of the image displacement. Therefore, if one increases the size of image displacement while holding pixel size constant, the maximum relative inaccuracy due to discretization must decrease. This can be seen in Figure 6.19, where the maximum possible orientation error decreases when larger image displacements are considered. It should be noted that, as seen in Figure 6.19, it may be the case that the error in orientation actually improves when shorter image displacements are considered. However, in general, the worst case performance will be significantly better (in terms of orientation error) as the size of the image displacement increases.

There is, of course, a tradeoff involved in obtaining more accurate orientation estimates. By looking at larger image displacements, one assumes that a sparser sampling of the image accurately reflects the nature of the scene. As described in the previous section (section 6.1.2), going to a sparser sampling in order to decrease errors due to discretization of the displacement vector, one increases the possibility for errors due to finite spatial resolution.

It should be pointed out that, because of the discrete nature of the problem, the exact effects of slight errors in orientation cannot be determined a priori. This follows from the fact that the algorithm bases its depth estimates on comparing the the spatial and temporal variation in image brightness recorded at in a finite region of the field of view. Because of inaccuracies in orientations of the displacement vectors, the spatial variation in image brightness may not accurately indicate the exact amount of temporal variation needed to indicate the expected image event (a single pixel displacement). The inaccuracy of the estimate given by the spatial variation is dependent on the error in orientation and the spatial distribution of brightness values in the scene in the neighborhood of the point in question. In general, since it is not unreasonable to assume that the distribution of brightness values in the image varies smoothly, the effects of small errors in the orientations of displacement vectors can be assumed to be negligible. This assumption has been confirmed experimentally. Depth values obtained in instances where the discrete approximations to displacement vectors must be inexact (the displacement vectors do not line up with the rows or columns in the imaging array) have the same (empirical) accuracy as those obtained in instances where the discrete approximations are quite good (the vectors line up with the rows and/or columns or the imaging array).

6.2   Uncertainty in the Camera Motion Parameters

This section details the effects of inaccurate knowledge of the camera motion parameters (camera motion can be specified by six parameters: three translation and three rotation). Recall from Chapter 5 that, in the context of the IGA algorithm, a depth estimate consists of a range of depth values in which an object may lie. This range is defined by the following:

$$dz_{i-1} \cdot r \cdot \cos\phi < d \leq dz_i \cdot r \cdot \cos\phi, \tag{6.9}$$

where ϕ is the angle between the camera's optical axis and the direction of motion, r is the distance (in image terms) between the point in question and the FOE, dz_i is the distance the camera displaced between acquiring the reference frame and the first frame in which the temporal gradient exceeded the spatial intensity gradient, and dz_{i-1} is the distance the sensor displaced between acquiring the reference frame and the previous frame. From this equation it is clear that inaccurate knowledge of the camera motion parameters will translate directly into inaccurate depth estimates.

Two scenarios are considered. The first is the "translation stage" scenario, where camera motion is assumed to be controlled by some device (such as a translation stage) whose precision is known. In this scenario, pure translational motion can be assumed. That is, in this scenario, there exists some coordinate system in which five of six of the camera motion parameters (two translation and three rotation parameters) can be assumed to be zero. Uncertainties are associated with the distance the camera displaces between acquiring frames (dz) and the orientation of the camera with respect to the axis of translation (ϕ). These uncertainties may arise due to inexact alignment of the camera on the translation device or mechanical limitations of the translation device. The second scenario under consideration is a worst-case scenario in which uncertainties are associated with all of the camera motion parameters. An example of this scenario would be using the motion of a vehicle to translate the sensor. Although the motion will be, for the most part, translational along one axis, there will be aberrations in this motion (both translational and rotational) due (among other things) to imperfections on the surface on which the vehicle is moving.

6.2.1   THE "TRANSLATION STAGE" SCENARIO

In the "translation stage" scenario, uncertainties are associated with the distance the camera displaces between acquiring frames (dz) and the orientation of the camera with respect to the axis of translation (ϕ). From equation 6.9, one can see that uncertainties in the value of dz will propagate linearly through the computations. Therefore, for example, if the knowledge of dz is accurate to ten percent, the boundaries of the depth range will be accurate to ten percent. The effects of inaccurate knowledge of dz were observed experimentally in some of the initial experiments performed using the IGA technique. In these experiments, the results obtained were consistently low estimates of the actual distances to objects in the scene. The cause of this turned out to be roundoff error in the controller used for the translation stage. Because of this error (in the conversion from millimeters of translation to steps of the motor), the translation stage was actually moving slightly more than the amount instructed. Once this was rectified, the depth values obtained were consistent with the actual locations of objects in the field of view.

The effects of uncertainty in the angle of orientation are not as obvious as those of uncertainty in the distance translated. Although, from equation 6.9, the value ϕ appears explicitly only in the cosine term, ϕ also plays a role in determining the value of r. Recall that r is the distance (in image terms) between the image location and the FOE. Since the location of the FOE is defined by the orientation of the camera ϕ, the value of r is dependent on the value of ϕ. From Chapter 3 we can express r as follows:

$$r = \sqrt{y^2 + f^2 \tan^2 \phi + 2fx \tan \phi + x^2}, \qquad (6.10)$$

where (x, y) is the location of the pixel in the image, and f is the focal length of the lens. From equations 6.9 and 6.10 one can conclude that an uncertainty of magnitude ξ in the estimate of ϕ would result in an uncertainty u in the depth estimate of:

$$u = \mid dz_i [r_{2_\phi} \cdot \cos \phi - r_{2_{\phi-\xi}} \cdot \cos(\phi - \xi)] \mid, \qquad (6.11)$$

where r_ϕ is given by equation 6.10 and $r_{\phi-\xi}$ is defined as follows:

$$r_{\phi-\xi} = \sqrt{y^2 + f^2 \tan^2(\phi - \xi) + 2fx \tan(\phi - \xi) + x^2}. \qquad (6.12)$$

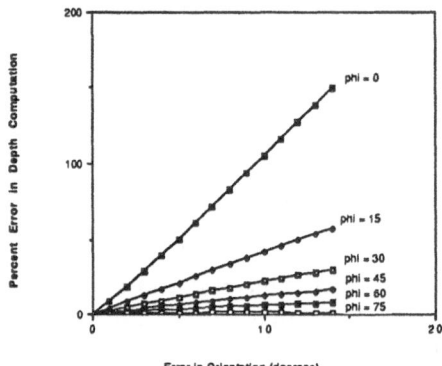

FIGURE 6.20. The absolute value of the percent error in the depth estimate versus ξ, the error in the orientation estimate. Plots for $\phi = 0, 15, 30, 45, 60, 75$ are shown ($dz = 1$ $f = 100$ and $(x, y) = (10, 10)$).

The effect of uncertainties in ϕ depend on the pixel location and the focal length of the lens. Plots of the absolute value of the percent error in the depth estimate versus ξ for several values of ϕ are given in Figure 6.20. Note that the uncertainty increases roughly linearly with ξ and that the technique is more sensitive to uncertainties in orientation for camera motion where the angle between the optical axis and the direction of motion is small.

A less obvious effect of uncertainty in the value of ϕ is the fact that an uncertainty in ϕ may cause the orientations of the displacement vectors to be incorrect (due to errors in the estimate of the location of the FOE). The effect of this uncertainty is similar to the effect of the errors in orientation arising from the discrete approximations to the actual displacement vectors. As described in section 6.1.3, because the distribution of brightness values in the neighborhood of a given point can be assumed to vary smoothly, the errors due to inexact estimates of the orientations of the displacement vectors can be assumed to degrade gracefully as the error in orientation increases.

6.2.2   THE WORST-CASE SCENARIO

In the worst-case scenario, it is assumed that the camera is being translated using some device which may introduce both unexpected translational components and unexpected rotational components to the camera motion. In general, it can be inferred that the IGA approach will be sensitive to such fluctuations in the camera motion parameters. This follows from the fact that IGA is looking for pixel-sized events to take place. Because a pixel

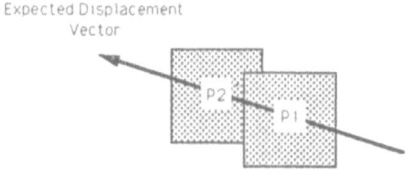

FIGURE 6.21. Pixel location P1, the expected orientation of the displacement vector at that location and P2, the next pixel location on the displacement vector.

corresponds to a very small region in the scene (approximately 0.12 × 0.12 degrees visual angle for a 512 × 512 imaging array and a 60 degree field of view), unexpected changes in sensor orientation greater than the extent of a pixel will, necessarily, induce unexpected variations in image brightness that are perceivable at the pixel level. These unexpected events may be interpreted incorrectly by the algorithm as corresponding to single pixel displacements, causing incorrect depth values to be reported.

The effects of these unexpected components of motion will be to alter the magnitude and orientation of the expected displacement vector at each pixel location in the image. Consider Figure 6.21. This shows a pixel location P1, the expected orientation of the displacement vector, and pixel location P2, the location on the displacement vector a fixed distance farther from the FOE. Assuming pixel location P1 corresponds to the location of a visual depth cue, and that the camera moves exactly as planned, when the image brightness at pixel location P2 equals that perceived at image location P1 in the first frame in the sequence, one can assume that the object giving rise to the visual depth cue at pixel location P1 in the reference image has displaced a fixed distance in the image. However, if there is an unexpected added component to the camera's motion, this may not be the case. Figure 6.22 shows the same scenario as before, but, in this case, the camera's motion has an unexpected added component. The projection on the image of the added motion component is oriented vertically, adding a downward component to the expected displacement vectors. This results in true displacement vectors that are oriented much differently than the expected ones. Because of this, the temporal variation at pixel location P2 may not accurately indicate the displacement of an object at location P1 in the reference image. In fact, in this case, one should monitor pixel location P2' to determine the behavior of an object at location P1 in the reference image. Unfortunately, there is no way of determining a priori that P2' is the location that should be monitored.

The problems associated with arbitrary uncertainty in the camera motion parameters are potentially much more significant than those previously

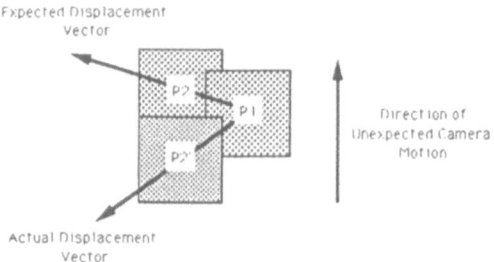

FIGURE 6.22. Pixel location P1, the expected orientation of the displacement vector at that location, P2, the next pixel location on the expected displacement vector, the actual displacement vector (cause by the unexpected camera motion) and P2', the next pixel location on the actual displacement vector.

discussed. Problems such as the temporal stability of the sensor can be modeled in such a way so that the worst-case behavior can be predicted. Similarly, problems due to limited spatial resolution and the geometry of the imaging array will cause the output of the algorithm to degrade in a known way. Even uncertainties in the camera motion can be modeled and dealt with nicely, if those uncertainties are limited to some simple cases (uncertainties in the magnitude of dz or small uncertainties in ϕ). However, when the camera motion cannot be controlled in such a manner so as the uncertainties in the orientation of the displacement vectors are kept within a reasonable limit ("reasonable limit" will be defined shortly), the behavior of the algorithm becomes unpredictable. Basically, whenever the orientation vector is altered in such a way so as the expected pixel location on the expected displacement vector does not accurately reflect the true pixel location on the actual displacement vector, the orientation of the displacement vector can be assumed to be off by more than the "reasonable limit" and the algorithm will break down. This limit, therefore, is a function of the effective pixel size and the expected image displacement. If the effective pixel size is large[2] and the expected image displacement is small (a fraction of the effective pixel size), there is a larger tolerance for error in the orientation of the displacement vector. However, if the effective pixel size is small and the expected image displacement is large (with respect to the effective pixel size), there is little tolerance for error.

There are, essentially, three ways of dealing with this problem: assume it's someone else's problem, add additional sensing capability, or use multiple sensors. The first solution is to do nothing at the IGA stage and assume

---

[2]The physical pixel size is defined by the resolution of the digitizer and the field of view of the camera. However, the effective pixel size is defined by the spatial extent of the smoothing filter used.

that the next level in the system is capable of assimilating (very) noisy data. By assuming some distribution for the fluctuations in camera motion, it is conceivable that one could obtain some reasonable depth estimates by integrating the output of the IGA algorithm over time. This approach suffers from two major drawbacks. First, this may impose a significant computational burden on the system as the output of many passes of the algorithm may be required to produce a single stable depth estimate. And, secondly, because the behavior of the algorithm is, essentially, unpredictable when the camera motion is unstable, in some cases it may be too much to assume that the output of the algorithm is stable enough to recover decent range estimates (even when results are integrated over time).

The second solution is to add additional sensing capability to the system. The basic assumption in this approach is that the sensor motion is assumed to be "piecewise stable." That is, the motion of the sensor is stable translational motion, with occasional aberrations. An example of this would be a camera attached to a mobile platform that is moving down a sidewalk. In this scenario, the motion of the sensor would be, essentially, pure translation, but when the platform navigates across the cracks between the blocks in the sidewalk, the smooth motion of the sensor would be disturbed. By detecting the aberrations in sensor motion, the images acquired during this stage may be ignored. Only those images acquired when the sensor motion is stable would be processed. This could be accomplished using some sort of inertial sensor, such as a gyroscope, to detect when the camera motion deviated from pure translation by more than a reasonable amount (see Figure 6.23).

The third approach is to use multiple sensors, rigidly mounted in a predetermined configuration. Recall that, given a sequence of n images acquired using a single sensor displacing through space, that same sequence can be acquired using n sensors fixed in space. This is illustrated in Figure 6.24. Although there is an increase in the amount of hardware required to build the system, the multiple-sensor approach is appealing for several reasons. First and foremost, it alleviates the need to move the sensor. With the multiple-sensor configuration, there is no uncertainty in the camera "motion" parameters. Because each sensor is mounted rigidly, the system can be calibrated precisely and, therefore, the relative locations of each of the sensors are known a priori. Secondly, the multiple-sensor configuration allows the system to accurately determine the locations of moving objects. This issue will be addressed in the following section.

6.3 Moving Objects

The world we live in is dynamic. Unfortunately, most vision algorithms (depth recovery algorithms in particular) are designed with only a static environment in mind. In the development of the IGA algorithm, a static

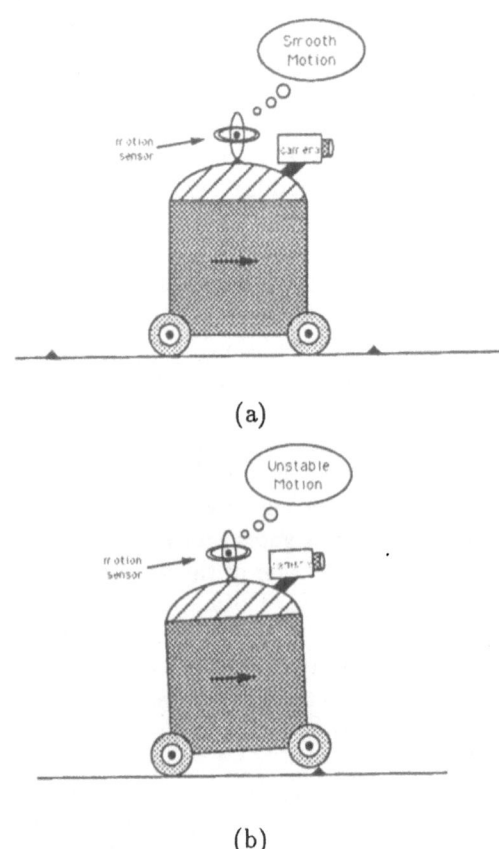

FIGURE 6.23. An inertial sensor can be used to detect deviations from smooth translational motion.

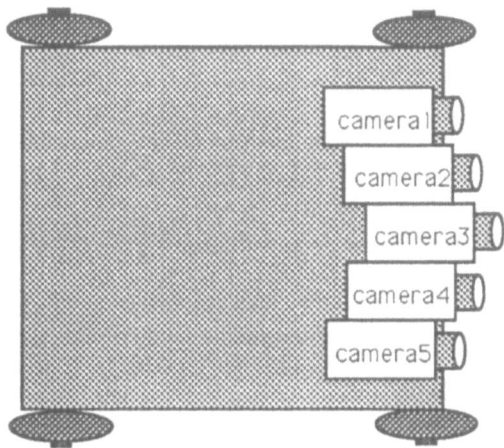

FIGURE 6.24. Multiple sensors rigidly mounted can be used to overcome problems with unstable camera motion.

environment was also assumed. This is because, in order to infer depth, it is necessary to assume that the temporal variation in brightness values is due to the image displacement induced by the sensor motion (see Chapter 4). Introducing moving objects invalidates this assumption. In an environment with moving objects, it cannot be assumed that the temporal variation in brightness values is due only to the image displacement induced by camera motion. Therefore, with moving objects, temporal variations in image brightness cannot be interpreted in the same manner as before.

Again, there are three ways of dealing with this issue: ignore moving objects and attempt to interpret the changes in image brightness as if all objects are stationary, add a pre-processing stage to isolate moving objects (and ignore those regions in the image where moving objects are identified), or modify the algorithm to use multiple sensors (this was alluded to in the previous section). Clearly the first alternative is unacceptable. Moving objects, in general, cause temporal variations in the brightness patterns in an image. Any attempt to interpret these changes as arising due only to sensor motion could potentially result in inaccurate range estimates at those locations in the image through which moving objects pass.

The second alternative is to add a pre-processing stage to isolate moving objects. Once these objects are isolated, the IGA algorithm would proceed as before, ignoring those locations in the field of view where moving objects are seen. To isolate moving objects, some sort of change detection algorithm (such as those proposed in [HNR84, JMN77, SJ89]) could be used on a sequence of images acquired from a fixed location in space. The image velocities of moving objects could be inferred by comparing the change detector outputs from each successive pair of images in the sequence. Once the image velocities are obtained, they can be used to predict how the moving objects will behave while the sensor is in motion. The sequence acquired while the sensor is in motion would be processed by the IGA algorithm as before, but those regions through which a moving object may have passed would be ignored.

This multi-stage approach has some serious limitations. First of all, it may not be reasonable to expect that the sensor motion can be controlled in such a manner so as sequences can be obtained both while the sensor is stationary and while it is moving. Consider an automotive application. It would certainly not be reasonable for the car to stop every ten feet, so a stationary sequence could be acquired. Secondly, the process of isolating moving objects and predicting their motion may be non-trivial in many cases. And, finally, in dynamic environment where multiple moving objects are present, the portion of the scene that is not ignored using the above approach may be too small to be of any use.

The third approach to the moving objects scenario is to use multiple sensors acquiring images simultaneously (instead of one sensor displacing through space). Consider the following: one way of accurately locating all objects in the field of view (including moving objects) would be to freeze all objects in space, translate a single sensor through the environment acquiring images along the way, then release the objects. The sequence of images acquired could then be used to accurately determine the locations of all objects in the field of view. Analogously, one could move the sensor infinitely fast, so that moving objects in the scene would appear stationary with respect to the sensor. Certainly, neither of these are viable alternatives. However, recall that, given a sequence of n images acquired in a static environment with a single moving sensor, an identical sequence can be acquired using n sensors fixed in space. If these sensors acquire their images at the same time instant, the sequence acquired would be identical to one acquired in the "make-believe" single sensor scenario described above. As stated in the previous section, this approach, although it requires more hardware to implement, is particularly appealing because, in addition to solving the problems imposed by moving objects, it eliminates uncertainties in the range estimates due to inexact knowledge of camera motion parameters. Note that the only modification needed in order for the IGA algorithm to work on a sequence of images acquired using multiple sensors is a semantic one. That is, the idea of a temporal variation in perceived brightness no longer appropriate. With multiple sensors, it is the inter-frame variation in perceived brightness that determines when a fixed pixel displacement has occurred.

6.4 Summary

In a real-world environment, the performance of the IGA algorithm can be expected to deviate from the ideal for three reasons: image sensors are not perfect, knowledge about camera motion may be inaccurate, and the assumption of static objects may not necessarily be valid. This section provides a summary of the analysis presented in the previous sections that addressed these issues.

Imperfect image sensors have poor temporal stability, limited spatial resolution and a fixed-geometry imaging array. To compensate for temporal instability, each image must be smoothed before it is processed by the algorithm. 4×4 or 8×8 averaging filters seem to adequately improve the temporal stability of the images, and a probabilistic model has been developed that indicates the worst-case performance of the system. There is no way to avoid errors due to the limited spatial resolution of the imaging device. However, it can be concluded that these errors must always occur on the "safe" side. That is, range errors due to limited spatial resolution will always indicate the objects being closer than they actually are, never farther. Finally, the fixed geometry of the imaging array of conventional devices causes inaccurate discrete approximations to continuous displacement vectors that do not align well with the imaging array. This effect can be reduced by looking for pixel displacements larger than the width of one pixel. In general, the larger the image displacement, the smaller the error due to discretization. However, looking for larger image displacements implies a reduction in the effective spatial resolution of the device, which leads to errors due to limited spatial resolution, as discussed above. In practice, it has been found that looking for image displacements on the order of the size of the smoothing filter used seems to work quite well.

Inaccurate knowledge of the camera motion parameters translates directly into inaccurate range estimates. If the inaccuracy in the camera motion parameters can be assumed to be limited to two special cases, uncertainty in the magnitude of translation (dz) and uncertainty in the orientation of the camera with respect to the axis of translation (ϕ), the uncertainties can be modeled nicely and incorporated into the range estimates. However, if the inaccuracy is not limited to these special cases, the performance of the algorithm becomes entirely scene-dependent, and therefore, unpredictable. One way to compensate for this is to add an additional sensing capability to the system that will detect when the sensor motion deviates from pure translation. Any images acquired during this phase should be ignored. The other way of dealing with this is to use multiple, rigidly mounted sensors. Using multiple sensors totally eliminates problems associated with uncertainties in the camera motion parameters because the sensors are mounted rigidly and can be calibrated precisely.

Moving objects cause temporal variations in the brightness pattern in that image that are not induced by the sensor motion. Therefore, any attempt to interpret them as such may result in inaccurate range estimates. One way of dealing with moving objects is to add a pre-processing stage that isolates moving objects in the field of view. Once the objects are isolated, those regions in the image through which they displace can be ignored. This approach assumes that it is possible to accurately identify the moving objects in the field of view – an assumption that may not be reasonable in many scenarios. A more reasonable approach is to use the multiple-sensor configuration proposed to deal with uncertainties in camera motion. If all the sensors acquired images at the same time instant, the sequence of images acquired would be analogous to one acquired by freezing the moving objects in space and using a single moving sensor acquire the sequence of images. This would result in accurate range estimates for all objects in the field of view, moving or stationary.

7

Fixed Disparity Surfaces

Because IGA looks for a specific type of image event to occur, it is possible to reason about the implications of that event before it occurs. More specifically, it is possible to determine for a predicted object location in what frame the object will first be perceived and the size of the depth range returned by the algorithm. Given this idea, it is possible to adjust the imaging geometry for a specific situation, tailoring the perceptual capabilities of the system to the specific needs of that situation.

In order to understand the implications of this idea more thoroughly, this chapter introduces the concept of a "Fixed Disparity Surface" (FDS). These surfaces are completely specified by the imaging geometry and the fixed disparity identified by the algorithm. FDS's can be used to effectively illustrate the perceptual capabilities of a given system configuration and the tradeoffs involved in the design of a depth recovery system. The first section of the chapter gives several examples of FDS's. The problem of how one interprets FDS's is discussed in the second section. Experimental results are presented that show how FDS's accurately reflect the perceptual capabilities of the IGA algorithm. The third section discusses how the FDS concept applies to conventional stereo.

7.1 Examples of FDS's

Because IGA looks for a single type of event, it is possible to draw a one to one correspondence between image points and 3D locations in the environment. This set of depth (d) values corresponding to each image location can be thought of as defining a "Fixed Disparity Surface" in three-space. For a particular FDS, the distance from the observer to a particular point on the surface is defined by the following equation (see Chapter 3):

$$d = dz \frac{r}{\delta} \cos \phi \tag{7.1}$$

where dz is the distance the camera displaced between acquiring frames, r is the distance (in the image) from the object to the FOE, δ is the (fixed) disparity that is to be detected, and ϕ is the angle of orientation of the camera with respect to the axis of translation.

Objects whose image displacement is exactly the fixed value must, by definition, lie on the FDS. Objects whose image displacement is less than the fixed value must lie beyond this surface (because depth is inversely

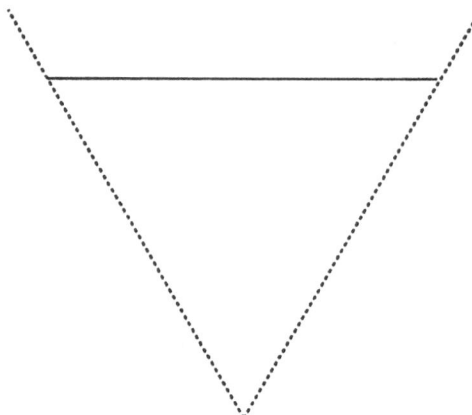

FIGURE 7.1. The FDS for an image pair acquired using $\phi = 90°$ (2-D intersection with the x-z plane). Dotted lines indicates field of view of camera.

proportional to image displacement) and, similarly, objects whose image displacement is greater than the fixed value must lie in front of this surface (between the observer and the surface).

Perhaps the simplest FDS's are those defined by lateral camera motion ($\phi = 90°$). In this case, the depth recovery equation becomes:

$$d = dz\frac{f}{\delta} \qquad (7.2)$$

where f is the focal length of the lens. These FDS's are planar and parallel to the image plane. Figure 7.1 shows a 2-D cross section of a FDS defined for $\phi = 90°$. The V-shaped dotted lines indicate the field of view of the camera; the camera is located at the "point" of the V.

At the other end of the spectrum are the FDS's defined by $\phi = 0°$ (axial camera motion). These surfaces are shaped roughly like the bell of a trumpet, circularly symmetric about the camera's optical axis (see Figure 7.2).

Examples of FDS's for $\phi = 15, 30, 45, 60$, and 75 degrees are shown in Figures 7.3 - 7.7. Note how the shape of the FDS's flattens out as ϕ grows larger.

7.2   Interpreting FDS's

Because it is the locations corresponding to fixed displacements that define the depth range in which an object lies, these "Fixed Disparity Surfaces" illustrate how the IGA algorithm will perceive the environment. Recall from Chapter 5 that if an object is first perceived in image i, it must lie

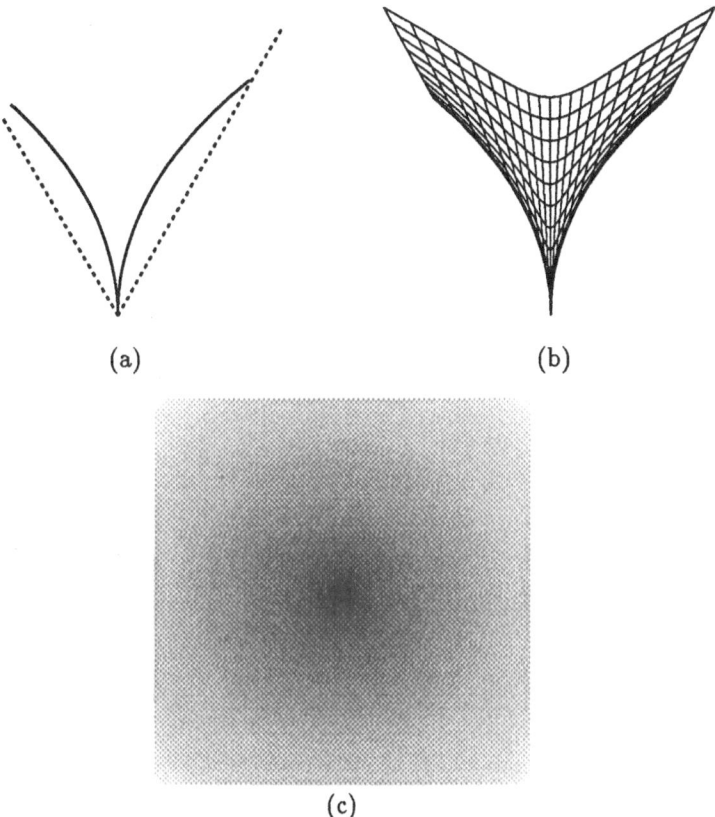

(a)   (b)

(c)

FIGURE 7.2. The FDS for an image pair acquired using $\phi = 0°$ (a) 2-D intersection with the x-z plane. (b) 2-D projection onto x-z plane. (c) "camera's eye
view," with distance to points on the FDS encoded as greylevels.

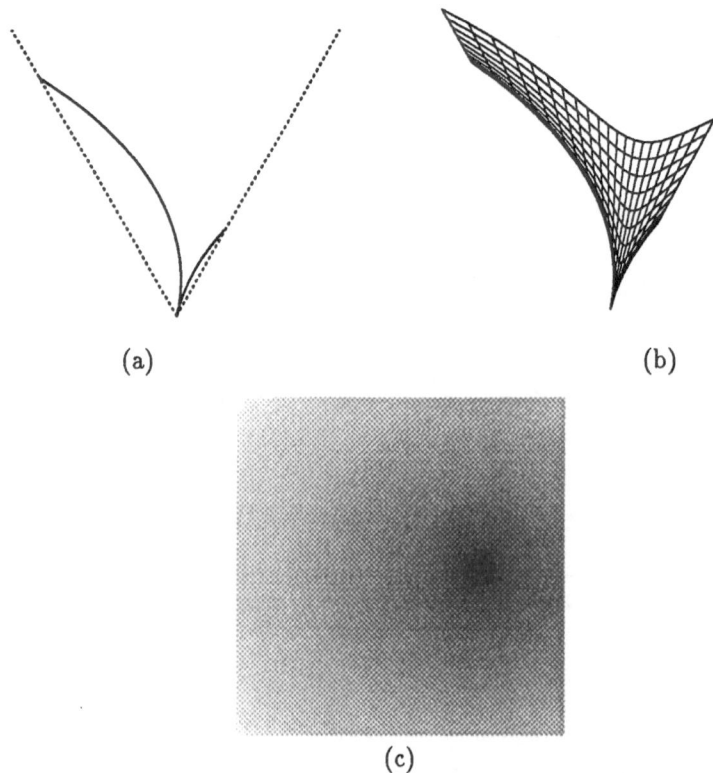

(a)   (b)

(c)

FIGURE 7.3. The FDS for an image pair acquired using $\phi = 15°$ (a) 2-D intersection with the x-z plane. (b) 2-D projection onto x-z plane. (c) "camera's eye view," with distance to points on the FDS encoded as greylevels.

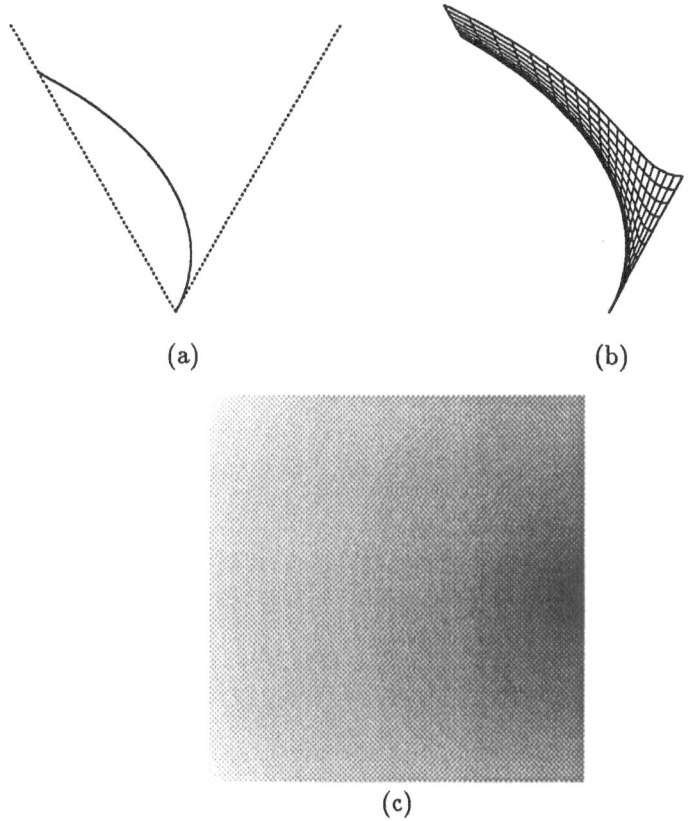

(a)   (b)

(c)

FIGURE 7.4. The FDS for an image pair acquired using $\phi = 30°$ (a) 2-D intersection with the x-z plane. (b) 2-D projection onto x-z plane. (c) "camera's eye view," with distance to points on the FDS encoded as greylevels.

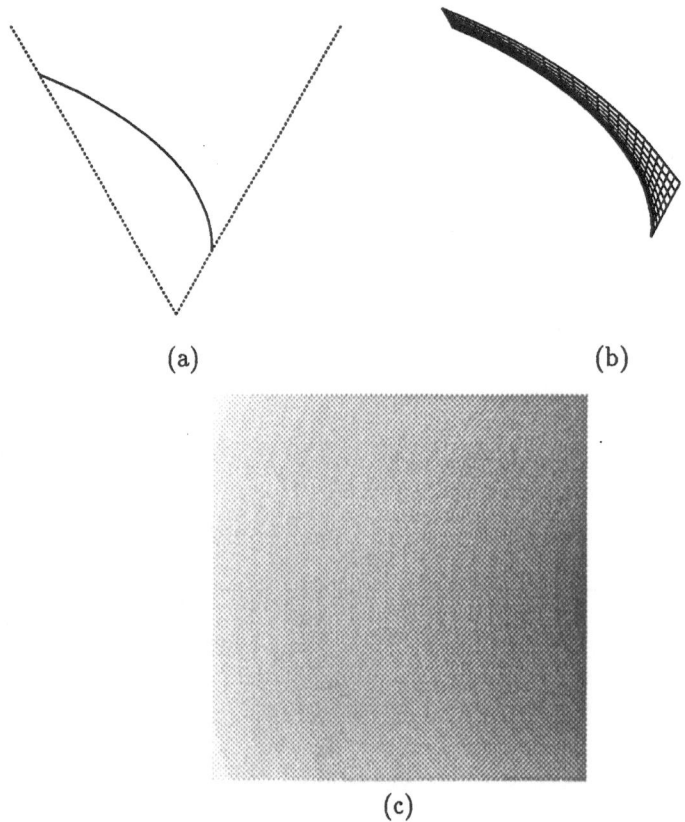

(a)   (b)

(c)

FIGURE 7.5. The FDS for an image pair acquired using $\phi = 45°$ (a) 2-D intersection with the x-z plane. (b) 2-D projection onto x-z plane. (c) "camera's eye view," with distance to points on the FDS encoded as greylevels.

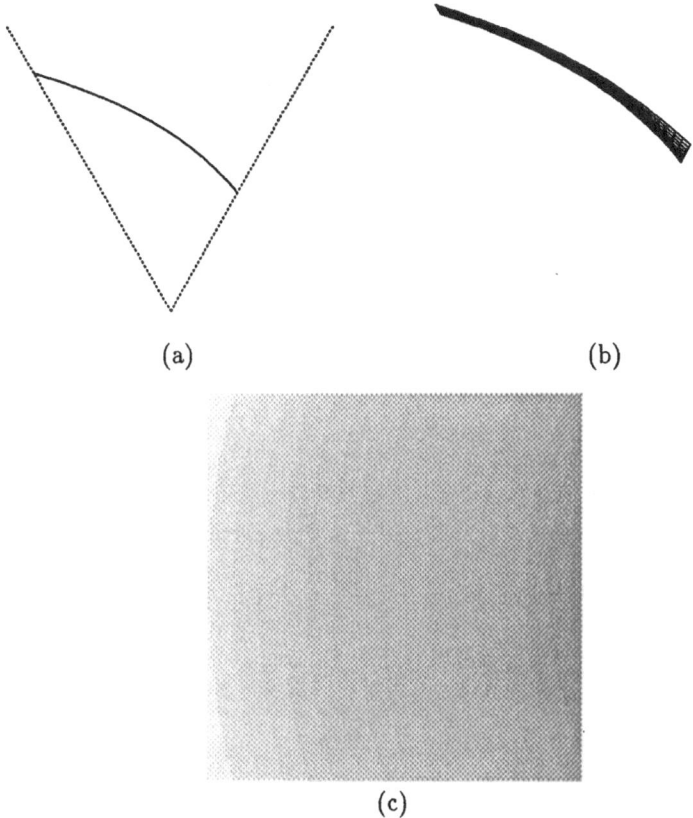

(a)   (b)

(c)

FIGURE 7.6. The FDS for an image pair acquired using $\phi = 60°$ (a) 2-D intersection with the x-z plane. (b) 2-D projection onto x-z plane. (c) "camera's eye view," with distance to points on the FDS encoded as greylevels.

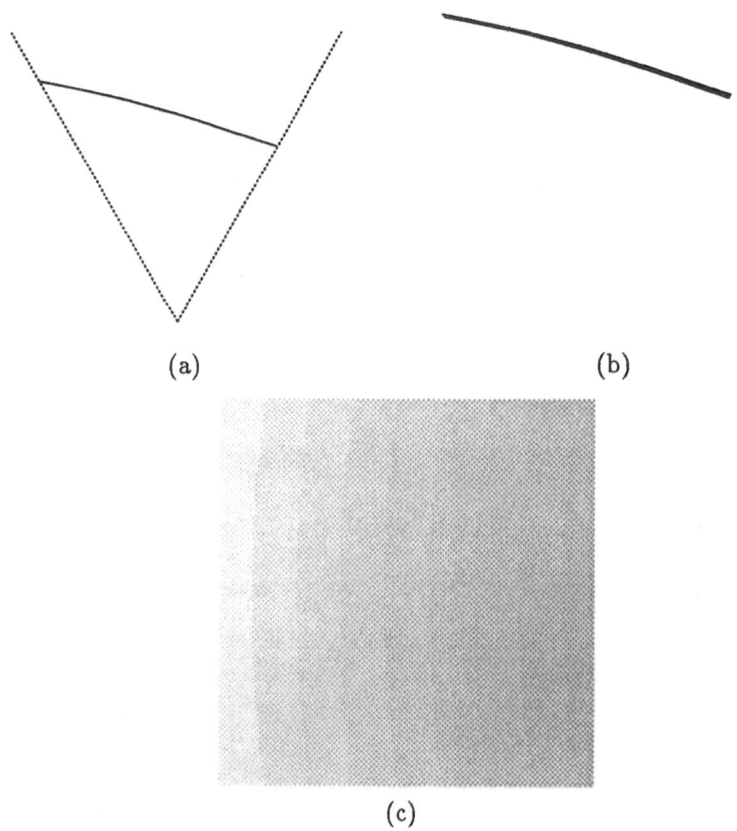

(a)   (b)

(c)

FIGURE 7.7. The FDS for an image pair acquired using $\phi = 75°$ (a) 2-D Inter-section with the x-z plane. (b) 2-D projection onto x-z plane. (c) "camera's eye view," with distance to points on the FDS encoded as greylevels.

FIGURE 7.8. A laboratory scene showing an object 160cm from the observer. The right edge of the object is 120 pixels from the center of the image.

in a region bounded by the depth corresponding to a fixed displacement in image $i-1$ and a fixed displacement in image i. This range corresponds to the distance between the FDS's defined by dz_{i-1} and dz_i at that location in the field of view. So, for a given set of images (and their associated dz's), the FDS's show how IGA "partitions" the environment. The nature of this partition is determined by the imaging geometry. The implications of this idea are that, for a given object location, one can determine a priori in which frame the object will first be perceived, and the size of the depth range in which the object must lie. This knowledge may be used to determine which imaging geometry may be best for a given application.

7.2.1   EXAMPLE: AXIAL CAMERA MOTION

Consider the following example. Given an imaging system which consists of a camera with a 480 x 480 imaging array and a lens with focal length equal to 695 pixels, suppose that there is an object in the environment located 160cm from the observer whose right edge appears 120 pixels from the center of the image. Figure 7.8 shows such a scene, and Figure 7.9 shows the corresponding map of the environment.

If a sequence of images is acquired using 1cm increments with axial camera motion and an image displacement of four pixels is to be detected, the object should first be perceived in the sixth frame (remember $d =$

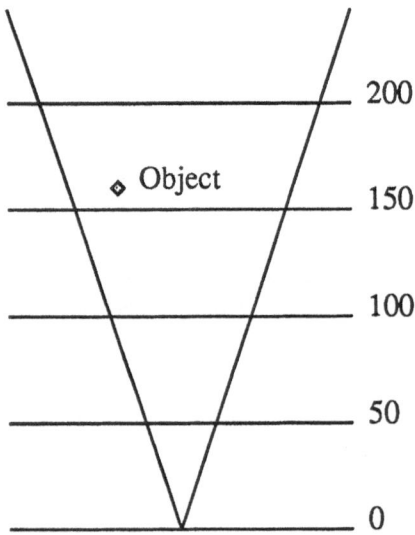

FIGURE 7.9. Example setup. Camera with focal length 695 and 480 x 480 resolution and an object 160cm from the observer. The V indicates the field of view of the camera and the horizontal lines indicated distance from the camera in centimeters.

$160cm$, $\phi = 0$ and $r = 120$) because:

$$5\frac{r}{4} < d \le 6\frac{r}{4} \qquad (7.3)$$

or,

$$150 < 160 \le 180. \qquad (7.4)$$

Figure 7.10 shows the FDS's associated with the given image sequence overlaid on a map of the environment. Note that the object is divided by the FDS corresponding to $dz = 5$. Figure 7.11 shows the output of the IGA algorithm when applied to such a sequence. IGA correctly identifies the right edge of the object as lying between the fifth and sixth FDS's and between 150 and 180cm from the observer. IGA also correctly identifies the left edge of the object as lying between the fourth and fifth FDS's and between 136-170cm from the observer (as predicted by the FDS's).

Recall that the depth range returned by the algorithm depends directly on the density of sampling of the image sequence (the size of the increments in dz), the location r of the object in the field of view, and ϕ, the angle of orientation with respect to the axis of translation. Clearly, in the above example, if 0.5cm spacing was used between frames, the object would be identified as lying in a range half the size of that identified with 1cm spacing (150 - 165cm as opposed to 150 - 180cm), or, if 4.0cm spacing was used, the range would be four times that determined by 1cm spacing (120 - 240cm). Similarly, if the object appeared 60 pixels from the FOE instead of 120, the object would, for 1cm sampling, appear first in the 11th frame, with an range of 15cm (150 - 165cm), instead of 30cm.

7.2.2 OTHER IMAGING GEOMETRIES

Although ϕ appears in the depth recovery equation (see Chapter 3) in only one place, the cosine term, it is implicit in the r term. Changing ϕ changes the location of the FOE and, thus, alters the distance from each image point to the FOE. For example, in the above scenario, if ϕ is changed to $15°$ (instead of $0°$), the FDS's would appear as shown in Figure 7.12. The right edge of the object appears first in the twelfth frame, with a depth range of 149 - 162cm (r would now be 56 instead of 120, due to the new location of the FOE). The output of the IGA algorithm as applied to this sequence is shown in Figure 7.13. Note that changing the sampling rate, or location in the field of view would have the same scaling effect as before.

Figures 7.14 through 7.18 show how the IGA algorthm partitions the environment using 1cm sampling and orientations of 30, 45, 60, 75, and 90 degrees, respectively. Table 7.1 summarizes the results of the experiments performed using different imaging geometries. Note that the experimental results exactly mirror the behavior indicated by the FDS's.

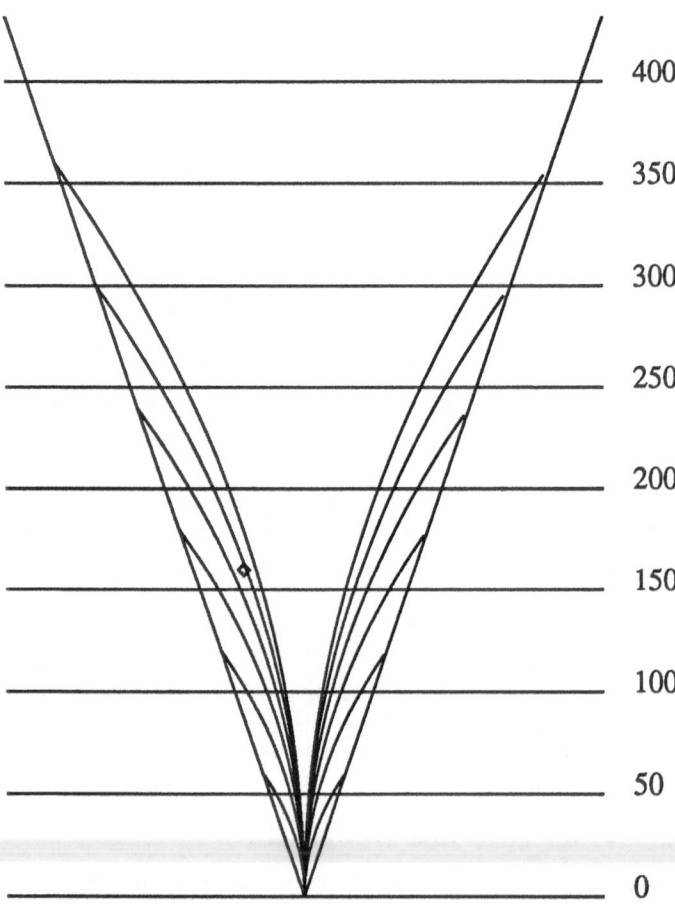

FIGURE 7.10. FDS's for axial camera motion, 1cm displacement between frames. The right edge of the object lies between the FDS's defined by $dz = 5$ and $dz = 6$.

FIGURE 7.11. The output of the IGA algorithm when applied to the axial motion sequence. Depth is encoded as greylevels, white indicates no information is present.

Experimental Results	
ϕ	perceived depth range
0	150 - 180
15	149 - 162
30	120 - 181
45	100 - 200
60	136 - 271
75	159 - 319
90	0 - 173

TABLE 7.1. Experimental results using different values of ϕ and 1cm sampling between frames. The actual object location is 160cm from the observer. Note that the object location in the image did not change, but, since the location of the FOE changes, r also changes.

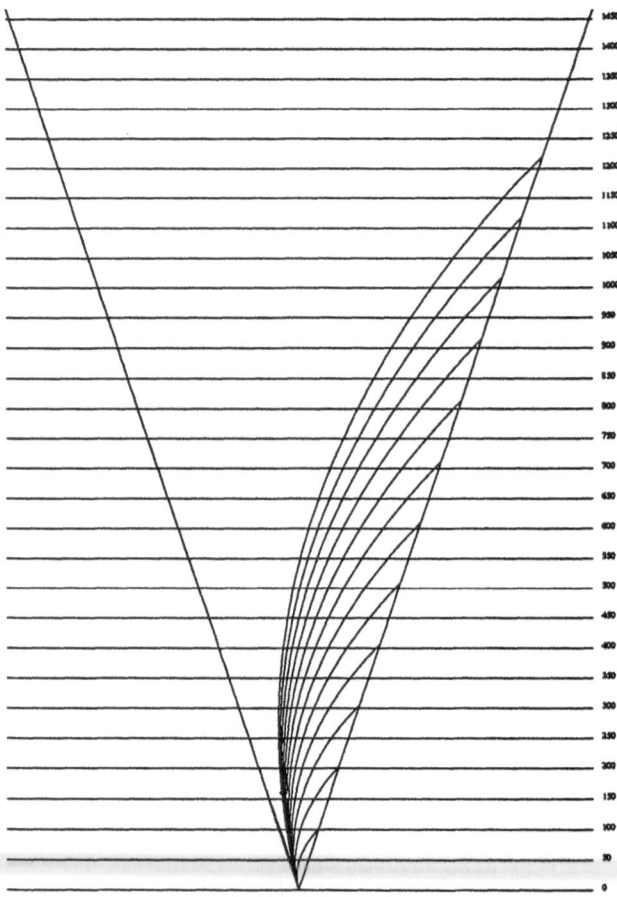

FIGURE 7.12. $\phi = 15°$, 1cm displacement between frames. The right edge of the object lies between the FDS's defined by $dz = 10$ and $dz = 11$.

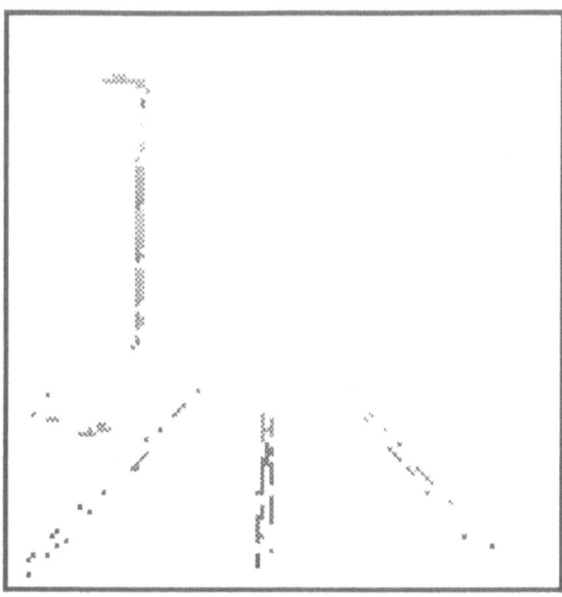

FIGURE 7.13. The output of the IGA algorithm when applied a sequence acquired using $\phi = 15$.

7.2.3 SELECTING AN IMAGING GEOMETRY

Perceptual needs vary from application to application. For example, a docking system may need a great deal of precision in the depth measurements in a certain area of the field of view (i.e. around the location of the docking port), while allowing greater ambiguity in other regions. Such an application would require that the FDS's be spaced close together near the point of interest. Another application, such as navigation, may allow for more ambiguity in the range estimates. In this case, the FDS's may be spaced farther apart. Certainly, imaging geometry plays a significant role in determining the perceptual capabilities of a system. FDS's can be used to determine what imaging geometry is best suited to a particular application.

For the most general of applications, sequences acquired with $\phi = 90°$ may be most appropriate. Recall that the FDS's corresponding to $\phi = 90°$ are planar and parallel to the image plane and that the depth range returned by the algorithm is independent of the location of the object in the image. Because $f > r \cos\phi$ for almost all (feasible) lens/image array combinations, $\phi = 90°$ gives one the most distance per dz. That is, for a sequence of n images, the distance from the observer to the FDS defined by $dz_n \cdot f$ is larger than the distance defined by $dz_n \cdot r \cdot \cos\phi$.

Camera orientations for which the FOE lies in the visible portion of the image plane may be inappropriate for many applications because locations near the FOE have small r values, and, therefore, require large dz values

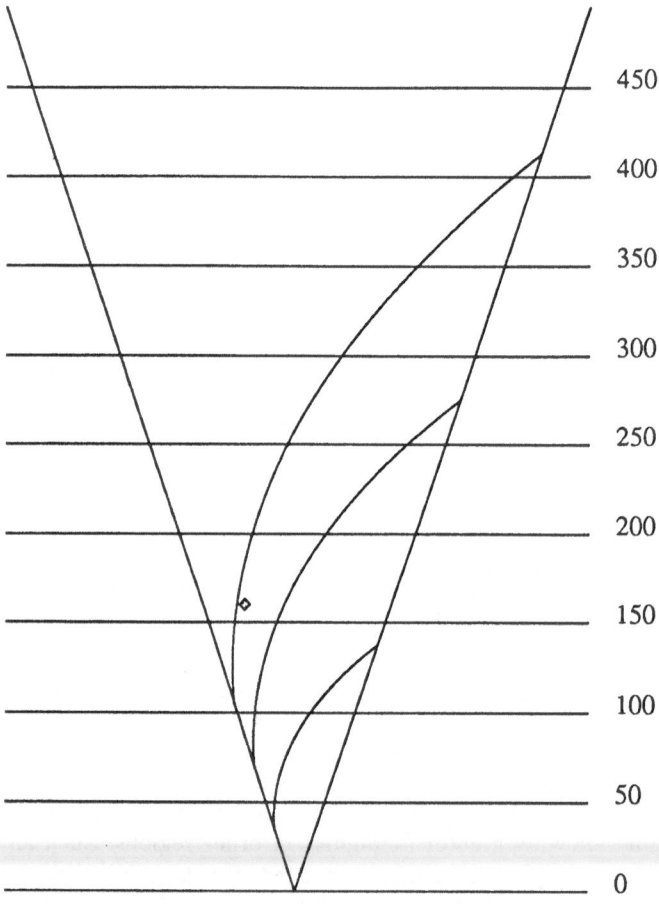

FIGURE 7.14. $\phi = 30°$, 1cm displacement between frames. The object lies between the FDS's defined by $dz = 2$ and $dz = 3$.

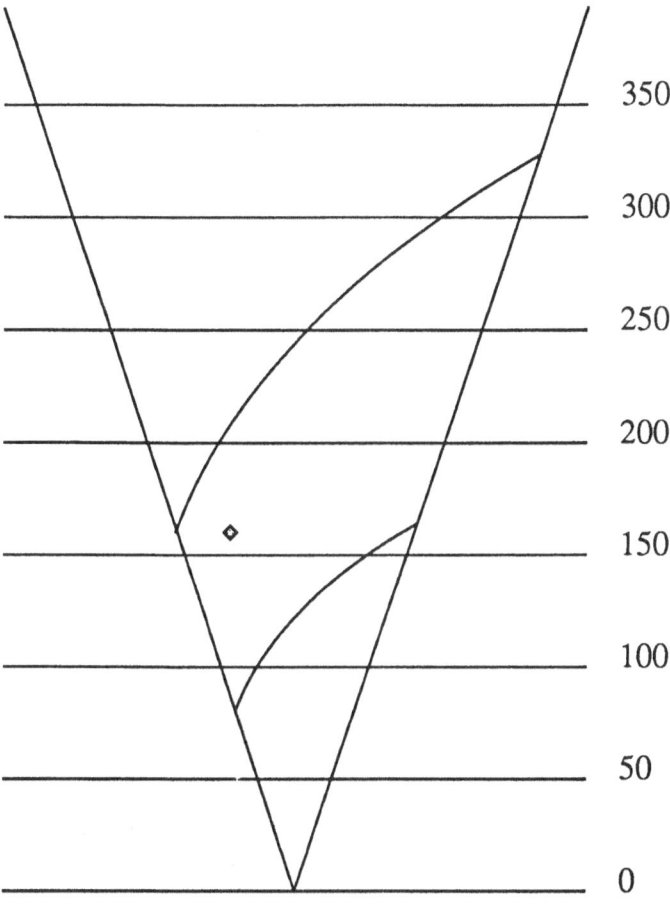

FIGURE 7.15. $\phi = 45°$, 1cm displacement between frames. The object lies between the FDS's defined by $dz = 1$ and $dz = 2$.

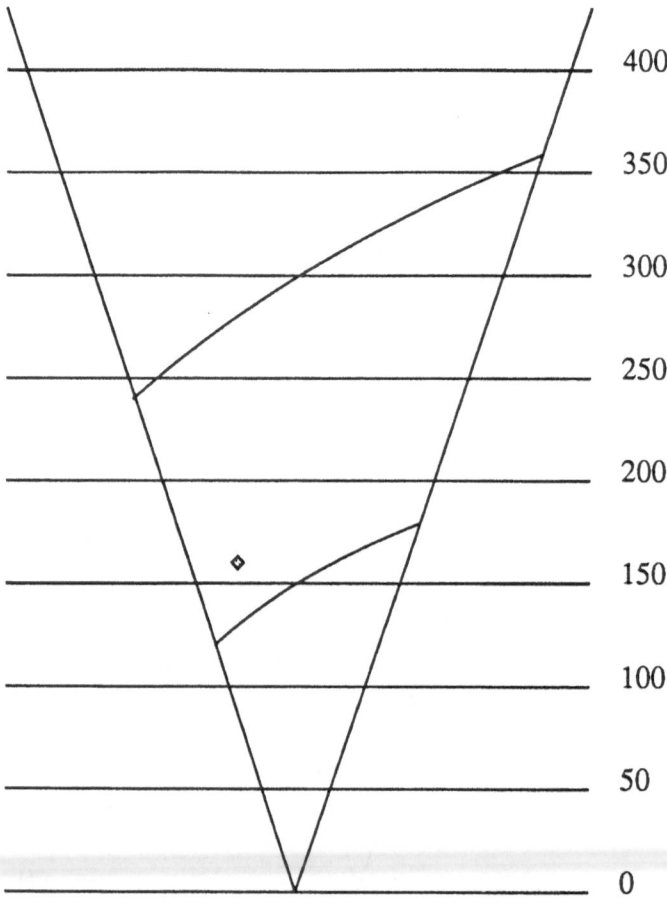

FIGURE 7.16. $\phi = 60°$, 1cm displacement between frames. The object lies between the FDS's defined by $dz = 1$ and $dz = 2$.

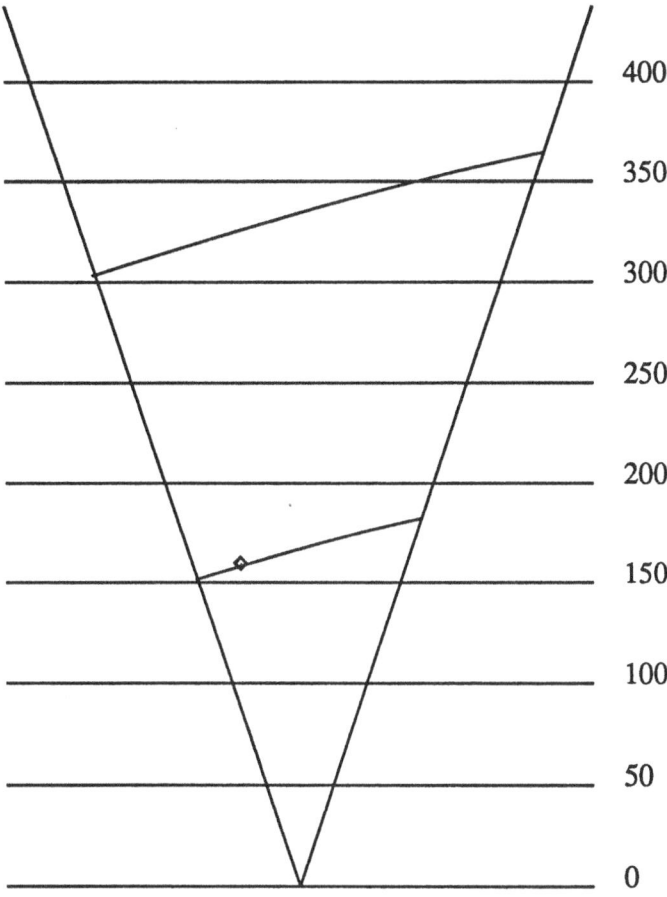

FIGURE 7.17. $\phi = 75°$, 1cm displacement between frames. The object lies between the FDS's defined by $dz = 1$ and $dz = 2$.

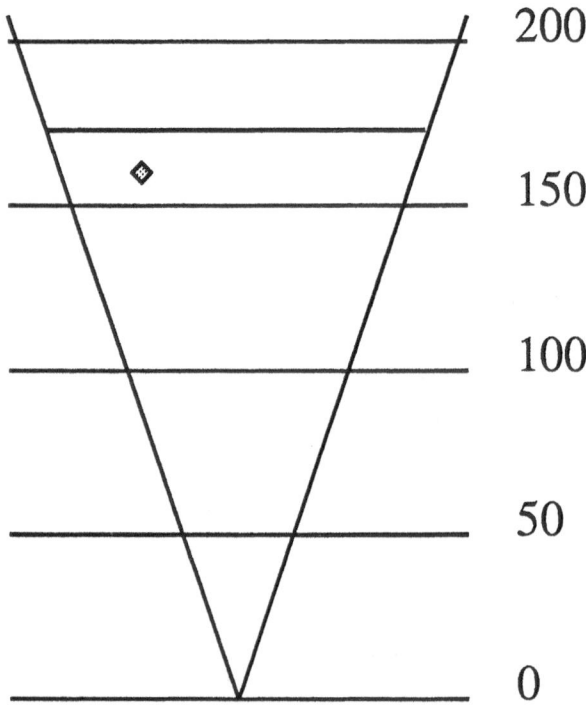

FIGURE 7.18. $\phi = 90°$, 1cm displacement between frames. The object lies between the FDS's defined by $dz = 0$ and $dz = 1$.

to perceive objects a reasonable distance from the observer. It may be appropriate to consider the region near the FOE as being a "blind spot", where little may be perceived.

One particularly appealing sensor geometry is the "V-shaped" setup shown in Figure 7.19. This arrangement has several advantages. Since this setup consists of, essentially, two different camera alignments (ϕ and $-\phi$), "blind spot" problems are avoided. This follows from the fact that the FDS's corresponding to $-\phi$ are the reflection (about the z axis) of the FDS's formed by ϕ, and, therefore, the blind spot of the FDS's corresponding to ϕ correspond to "good" regions of the FDS formed by $-\phi$. Also, in addition to the two sequences of images (corresponding to cameras 0 to 2 and 2 to 4 in this example) one can also consider all possible combinations of two images (i.e. 0-3, 0-4, 1-3, etc...) and obtain additional information. That is, instead of just four FDS's (0-1,0-2,4-3,4-2), one would have ten (0-1,0-2,4-3,4-2,0-3,0-4,1-3,1-4,2-3,2-4). The FDS's corresponding to the imaging geometry shown in Figure 7.19 are illustrated in Figure 7.20.

For example, consider the scenario described in section 7.2.1, where an object appears in the left hand side of the image, 160cm from the observer and 120 pixels from the image center. If a camera orientation of $\phi = 45°$ is used and images are acquired at 1cm intervals, IGA will perceive the object as lying in a range of 100 - 200cm from the observer (see table 7.1). If a camera orientation of $\phi = -45°$ is used, IGA will perceive the object as lying in a range of 143 - 287cm from the observer. Combining these two results, one can conclude that the object must lie in a range of 143 - 200 cm from the observer.

Adding more sensors to the same geometrical configuration will serve to either increase the effective range of the system (if the spacing between sensors is kept the same), or decrease the size of the depth range if the spacing is reduced (because a finer partition will be induced on the environment). Figures 7.21 and 7.22 show the previous imaging configuration and its associated set of FDS's for eleven sensors, instead of five.

7.3 Fixed Disparity Surfaces and Conventional Stereo

An analogy can be made between FDS's associated with each image in a sequence used by the IGA algorithm and the surfaces defined by different disparities for a conventional binocular stereo setup. For a given binocular pair, the surfaces defining the perceptual capabilities of the system are obtained by looking at different disparities and a fixed camera displacement (instead of different camera displacements and fixed disparities). The uncertainty associated with a given depth estimate, in the binocular stereo case, is dependent on the localization properties of the feature being used

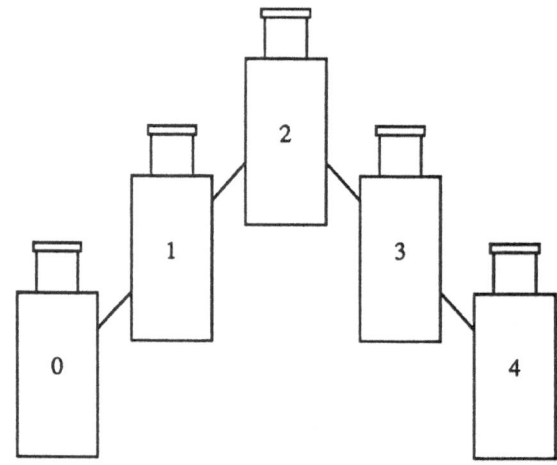

FIGURE 7.19. V-shaped sensor geometry. This particular example consists of five cameras (or sensor locations), with cameras 0-2 arranged with orientation $\phi = 45°$ and cameras 2-4 oriented with $\phi = -45°$.

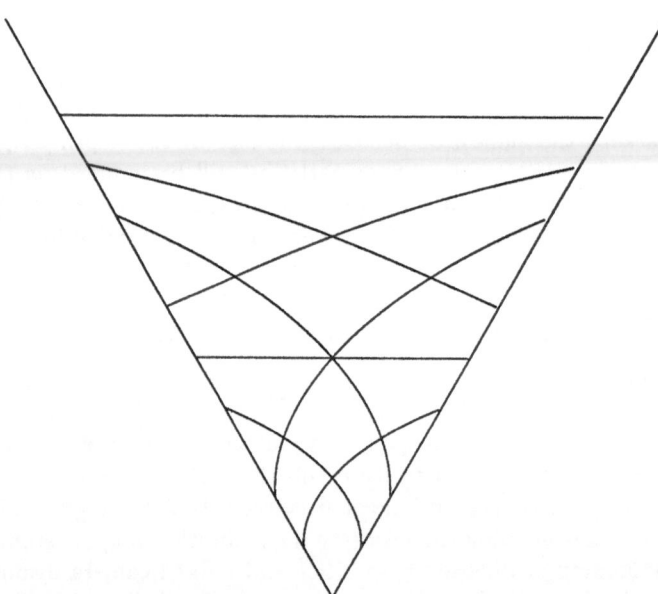

FIGURE 7.20. The FDS's formed by a V-shaped sensor geometry using a total of five cameras at orientation $\phi = 45°$ and $\phi = -45°$.

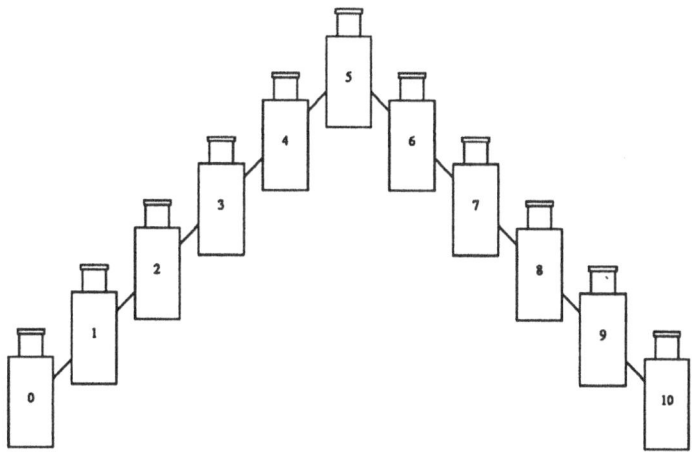

FIGURE 7.21. V-shaped sensor geometry. This particular example consists of eleven sensors.

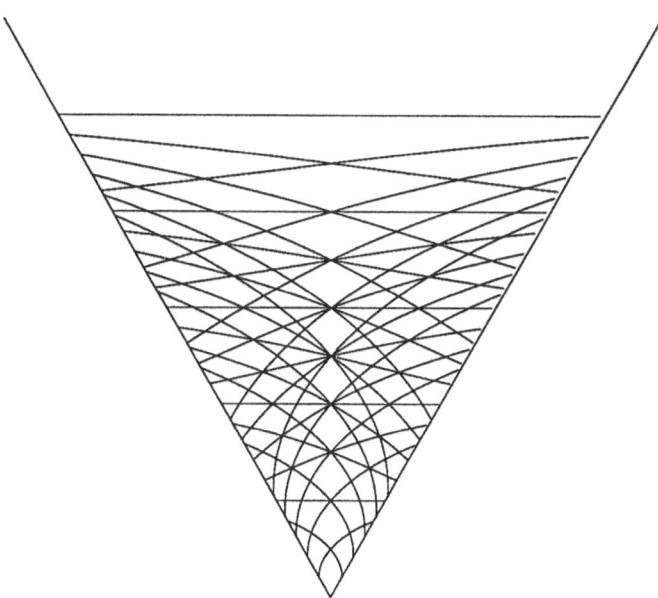

FIGURE 7.22. The FDS's formed by a V-shaped sensor geometry using a total of eleven sensors.

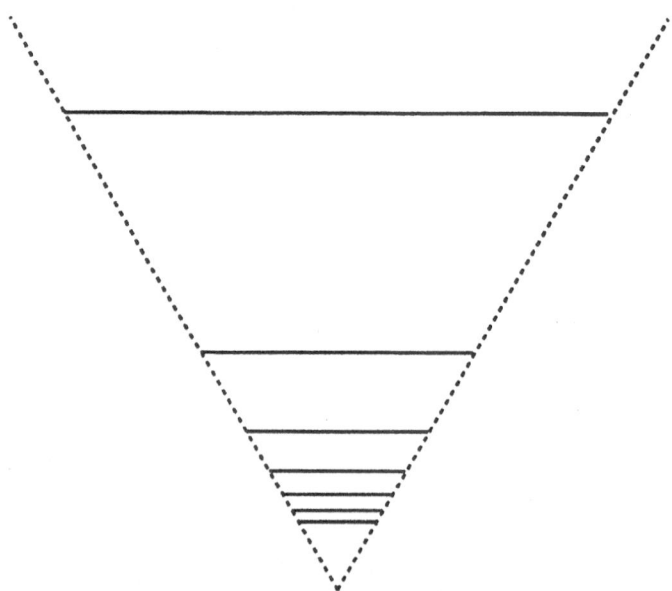

FIGURE 7.23. The Fixed Disparity Surfaces formed by a binocular stereo image pair acquired in such a manner that the optical axes of the two cameras are parallel. Dashed lines indicate the field of view of the camera, solid lines indicate the FDS's.

for matching. For example, if there is a one pixel uncertainty in the location of the features, a disparity of x implies only that the disparity lies between $x - 0.5$ and $x + 0.5$. Therefore, the Fixed Disparity Surfaces corresponding to $\delta = 0.5, 1.5, 2.5$, etc ... would partition the environment in the same manner as the FDS's used with IGA, indicating the region in space where objects may lie, given that they are perceived with a particular disparity. Figure 7.23 shows the fixed disparity surfaces corresponding to a binocular stereo pair acquired with cameras oriented so that their optical axes are parallel. Note that, as expected, the ambiguity associated with a given depth estimate increases as the perceived disparity decreases.

8

Experiments

The purpose of this chapter is threefold: to describe the setup used to perform experiments using the IGA algorithm, to describe the procedure used to calibrate the equipment, and to describe several experiments performed using IGA. There are several parameters associated with the IGA algorithm; the experiments discussed in this chapter illustrate the implications and tradeoffs involved in selecting different parameter values. Additionally, these experiments illustrate the performance of IGA in several different imaging scenarios.

8.1   Experimental Setup

The equipment used in the majority of the experiments consisted of a single Sony XC-77 video camera, a pan/tilt/translation (*ptt*) stage to move the camera (assembled from Velmex components), and a Silicon Graphics 4D-210GTX workstation with the live-video digitizer option. An Imaging Technologies digitizer board mounted in an Apollo DN3500 workstation was also used for many of the experiments. Figure 8.1 shows the Sony camera mounted on the *ptt* stage. The translation portion of the *ptt* stage employs a lead screw with an accuracy of better than 0.04mm / 30cm [Vel], or 0.013%. The *ptt* head is controlled using a Velmex model 8313 three-axis controller, which receives its commands from either the Apollo or Silicon Graphics workstation.

The majority of the experiments were performed in the mobile robot workspace at the University of Michigan Artificial Intelligence Laboratory. This workspace consists of a 5 x 6 meter open floorspace on which is marked a 20 centimeter grid. The grid markings are accurate to about one centimeter. Figure 8.2 shows the mobile robot workspace.

8.2   Calibration Procedures

Two procedures must be performed before experiments can be run using this equipment. Each procedure is described in detail in a following section. The first step is to align the camera properly on the *ptt* stage, which ensures that the motion specified by the controller corresponds to the actual motion of the camera. The second step is to determine the focal length of the imaging system. This is needed for the depth computation, and is obtained experimentally.

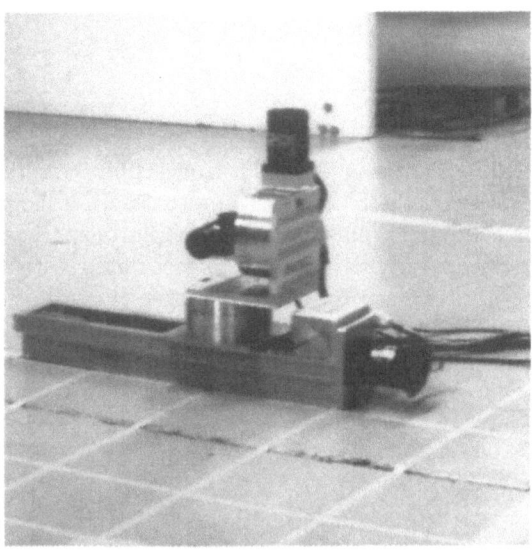

FIGURE 8.1. The Sony camera mounted on the pan/tilt/translation stage.

FIGURE 8.2. The mobile robot workspace at the University of Michigan Artificial Intelligence Laboratory.

8.2.1   CAMERA ALIGNMENT

The first procedure is to align the camera with the axes of the *ptt* stage. This is initially accomplished by simple line of sight techniques. That is, the camera is mounted on the *ptt* stage so that it appears to line up correctly with the axes of the *ptt*. Then, the camera is aligned using an iterative process in which the camera is moved (using the translation portion of the *ptt*), and the pan and tilt zeros are adjusted until accurate alignment is accomplished. The following is a summary of this procedure:

1. Orient the camera along the axis of translation.

2. Acquire and display an image, placing a dot in the center of the acquired frame.

3. Translate the camera, and acquire and display a second frame, again placing a dot in the center of the frame.

4. If the dot appears to move to the left of the target between frames, the FOE is located to the left, so modify the pan zero point by panning left some small amount. If the dot appears to move to the right, pan right.

5. If the dot appears to move towards the top of the target between frames, the FOE is located towards the top, so modify the tilt zero point by tilting up some small amount. If the dot appears to move towards the bottom, pan down.

6. If the dot doesn't move, the camera is aligned properly, otherwise GOTO 2.

Figure 8.3 shows the target used to align the camera.

This technique assumes that the optical axis of the camera coincides with the center of the digitized image. This assumption, of course, is true only if the camera lens is perfectly aligned with the receptor array. And, certainly, this alignment is good only to the resolution of the imaging device. Nevertheless, this technique has proven adequate for the purposes of this research. It certainly may be possible to align the camera more precisely using other more intricate techniques, however these are considered beyond the scope of this research.

8.2.2   DETERMINATION OF FOCAL LENGTH

The second step in calibrating the imaging system is to determine the focal length of the camera's optical system. The focal length is determined experimentally, using known camera motion perpendicular to the camera's

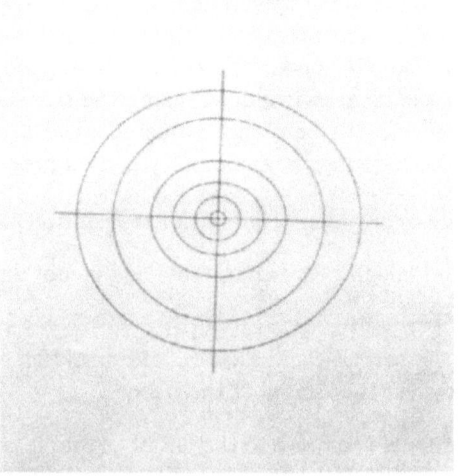

FIGURE 8.3. The target used to align the camera with respect to the pan/tilt/translation stage

optical axis and the relationship derived in Chapter 3 for recovering depth using lateral sensor motion:

$$d = \frac{dz \cdot f}{\delta},$$ (8.1)

where d is the distance to the target (measured along the camera's optical axis), dz is the baseline, or distance the camera traveled (perpendicular to its optical axis) between frames, f is the focal length of the lens, and δ is the disparity, or distance (in image terms) the target displaced between frames.

Note that, given the above equation, it is convenient to express the focal length in terms of pixel widths. Certainly it is possible to express focal length in terms of millimeters (or some other distance metric), if the pixel size is known. However, since the units cancel out anyway, the conversion from pixel widths to millimeters would only add additional unnecessary effort (and uncertainty) to the focal length calculation.

For example, Figure 8.4 shows the first 480 x 480 frame from the image pair used to determine the focal length of an 8.5mm Cosmicar lens. The target is located one meter from the camera, and a baseline of 20cm was used. Disparities were measured manually for the seven points on the target surface and the average disparity was computed to be 139 pixels. Using equation 8.1 with $d = 100cm$ and $b = 20cm$, the focal length was determined to be 695 pixel widths.

FIGURE 8.4. The first frame of the image pair used to determine the focal length of the lens. The determination of the focal length was based on disparities measured manually for the seven points on the target.

It should be noted that there are more precise techniques for determining the optical parameters of a system, but the above is sufficient for our needs. This follows from the fact that, in general, the reference values obtained for target locations are only as accurate as the calibrations on the floor in the mobile robot workspace. Any accuracy greater than that of the mobile robot workspace would not be perceivable.

8.3   Experiments

This section describes several experiments performed using the IGA algorithm. Essentially, there are seven variables or parameters that have a direct impact on the performance of an IGA-based depth recovery system. These parameters are:

- the size of the smoothing window used (sw),

- the threshold used to determine if a point is to be considered for depth recovery (T).

- the size of the disparity detected (δ),

- the camera spacing between frames (dz),

- the angle of orientation of the camera's optical axis with respect to the axis of translation (ϕ),

- the focal length of the camera lens (f),

- the degree of subsampling[1] (ss),

The following sections describe the results of several experiments that illustrate the tradeoffs and implications of selecting different parameter values. Of course, each of the parameter settings is not entirely independent of the other parameter values. Unfortunately, however, it would be impossible to illustrate the performance of IGA under all possible combinations of these parameter values. It may be the case that a particular application domain may require different values than those indicated in the following discussion. However, the intent here is purely to illustrate the performance of IGA in different imaging scenarios. Certainly, it may be the case that the performance of IGA in some particular application domain may differ slightly from that indicated in the following discussion.

8.3.1   SMOOTHING WINDOW SIZE

As shown in Chapter 6, because of the poor temporal stability of conventional imaging devices, it is necessary to smooth each image before it is processed. As also shown in Chapter 6, the temporal stability of the smoothed images improves as the size of the smoothing filter increases. Unfortunately, however, this does not imply that it is best to employ huge filters for the smoothing task. As filter size increases, so does processing time and, perhaps more importantly, as filter size increases, features (i.e. visual depth cues) become more and more blurred, resulting in fewer and fewer visual depth cues. For the purposes of this research, a simple averaging filter is used for the smoothing task.

Figure 8.5 shows the scene used to test the effects of smoothing window size on the performance of IGA. In this scene, the NEC box is 200cm from the observer, the chair is 300cm from the observer, the Silicon Graphics box is 460cm from the observer, the white box is 560cm from the observer and the partition and back wall are 700cm from the observer. For the purposes of these experiments, ten images were acquired at 1cm intervals $(dz = 1cm)$ using camera motion perpendicular to the camera's optical axis $(\phi = 90°)$. A disparity of 4 pixels was detected $(\delta = 4)$ and only those image locations whose spatial intensity gradient was greater than 8 greylevels were considered for the depth determination process $(T = 8)$. A

---

[1]The degree of subsampling indicates the ratio of pixels from the full resolution image used for computing the depth map (i.e. $ss = 4$ implies that one-fourth of the pixels are used, $ss = 8$ implies that one-eighth are used, and so on)

FIGURE 8.5. The scene used to illustrate the effects of altering the size of the smoothing window on the performance of IGA.

Cosimicar 8.5mm lens with a focal length of 695 pixels was used and no subsampling was done ($f = 695$ and $ss = 1$).

The algorithm was run on the above scene using a 2×2, 4×4, 8×8, 16×16, and 32×32 smoothing window, as well as no smoothing at all. Figures 8.6 (a) - (f) show the reference image from each of the sequences processed. Note that as the smoothing window size increases, the number of visual depth cues in the image decreases. Figures 8.7 (a) - (f) show the output of the IGA algorithm for each of sequences processed. Depth values are encoded as greylevels, with white implying no information is recovered. In this particular example, a greylevel of 20 implies a depth of 20 decimeters, a greylevel of 50 indicates a depth of 50dm, and so on.

For the no smoothing case (Figure 8.8 (a)), the output of IGA looks something like an edge detector, with the depth values recovered located at the intensity edges in the scene. As the size of the smoothing window increases (Figure 8.8 (b) - (f)), points corresponding to the locations of visual depth cues become blurred into regions. For the smaller smoothing windows, these regions continue to correspond to visual depth cues. Eventually, however, after the smoothing window becomes large enough, these regions become roughly uniform in intensity, and, therefore no longer correspond to visual depth cues. In general, as the size of the smoothing window increases, fewer and fewer image locations are identified as being candidates for depth recovery.

Figures 8.8 (a) - (f) show a subwindow of the output of IGA for each of the

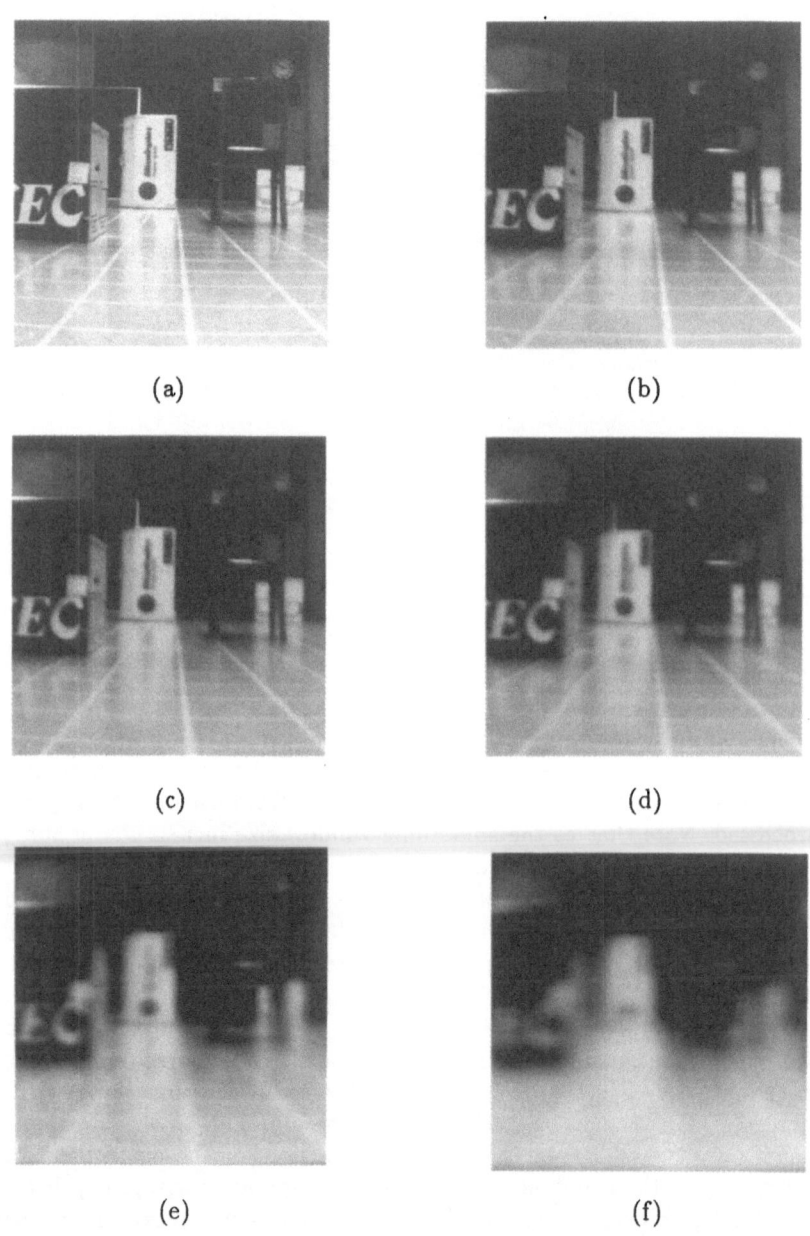

(a)   (b)

(c)   (d)

(e)   (f)

FIGURE 8.6. The reference images from each of the sequences processed. (a) no smoothing. (b) 2×2. (c) 4×4. (d) 8×8. (e) 16×16. (f) 32×32.

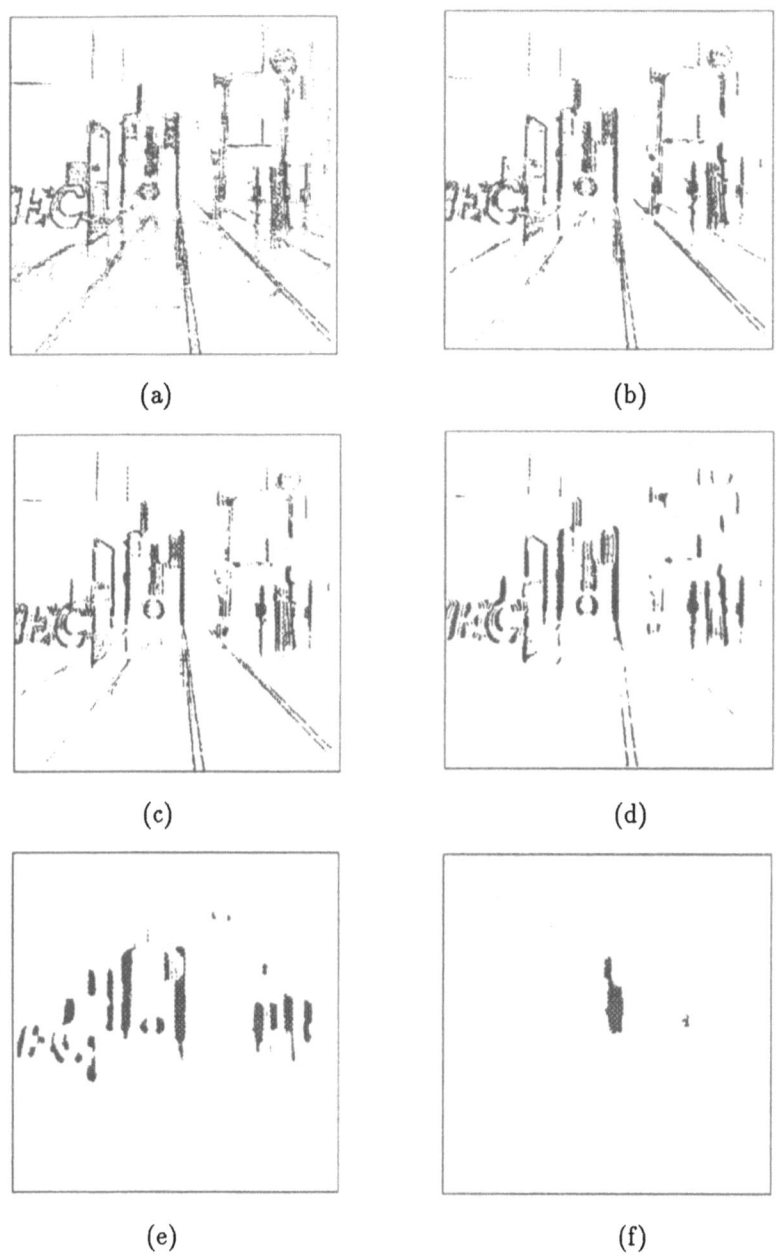

FIGURE 8.7. The output from the IGA algorithm for each of the sequences processed. (a) no smoothing. (b) 2 × 2. (c) 4 × 4. (d) 8 × 8. (e) 16 × 16. (f) 32 × 32.

above sequences. The region shown corresponds to the NEC box located in the left hand side of the scene. Note that in Figures 8.8 (a) and (b), corresponding to the no smoothing case and the 2 × 2 smoothing window case, the definition is very good – depth values are reported exactly at the locations in the scene where the depth cues exist, however, because of the poor temporal stability of the sensor, these values are quite inconsistent. For the 4 × 4 and 8 × 8 smoothing window cases, illustrated in Figures 8.8 (b) and (c), the depth values begin to be clustered in groups about the visual depth cues in the scene (due to the smoothing), but the depth values reported are much more consistent (and correct) than in the previous two cases. For smoothing windows greater than 8 × 8, the performance of the algorithm begins to degrade, as the visual depth cues in the reference image lose their definition. Figures 8.8 (e) and (f) show the output for the 16 × 16 and 32 × 32 smoothing window cases, respectively. Note that in the 32 × 32 case, no depth values are reported for this region at all.

Recall that, in the case of these experiments, lateral camera motion was used with 1cm spacing between frames. Given that the lens used has a focal length of 695 pixels and a disparity of 4 pixels was detected, this implies that the IGA algorithm will partition the environment in 173.75cm segments (see Chapter 7). Empirically, it has been shown that the 8 × 8 smoothing window is most appropriate for the types of scenes IGA has been tested on and the experimental setup used in these tests. The output shown in Figure 8.8 (d) correctly indicates the majority of the box as lying between 173 and 346cm from the observer, but several points are indicated as lying between 0 and 173cm from the observer. Fortunately, these incorrect indications are easily identified and removed using a simple smoothing algorithm. This smoothing algorithm uses a simple voting scheme to identify and reject depth estimates that are not consistent with their neighbors. The output of IGA before and after this smoothing has been performed is shown in Figure 8.9. After smoothing is done, all locations correctly indicate the box as lying between 173 and 346cm from the observer.

8.3.2 THRESHOLD VALUE

Ideally, all locations in the scene corresponding to non-zero spatial intensity gradients in the reference image should correspond to image locations where depth information can be recovered reliably. Unfortunately, this is only true if the sensors used have perfect temporal stability. As was shown in Chapter 6, this is clearly not the case with CCD devices, and, even though the input to the IGA algorithm is first passed through a smoothing filter, there is still no guarantee of perfect temporal stability. Therefore only those image locations whose spatial intensity gradient is greater than some threshold should be considered for the depth recovery process. This threshold should be large enough so that only those points that truly correspond to visual depth cues are considered, but the threshold should also be small enough so

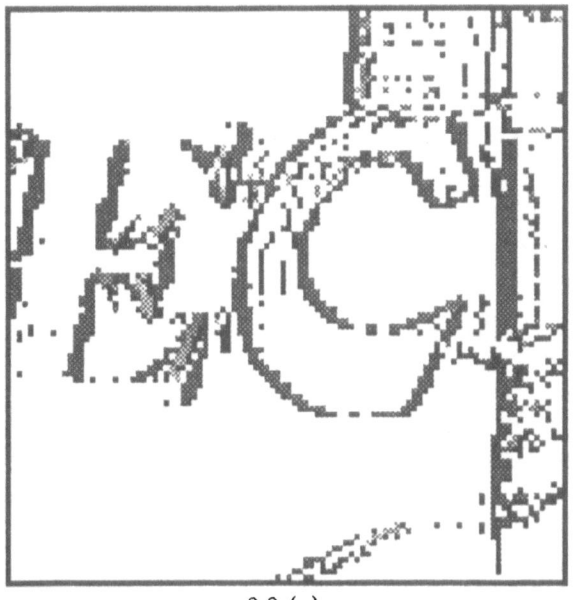

8.8 (a)

8.8 (b)

8.8 (c)

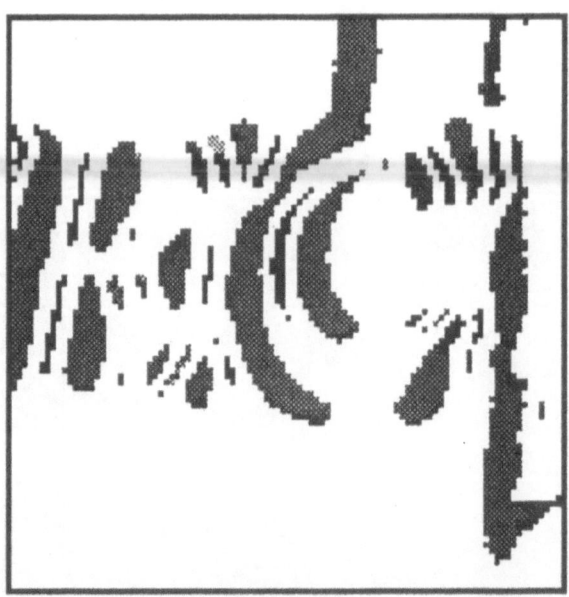

8.8 (d)

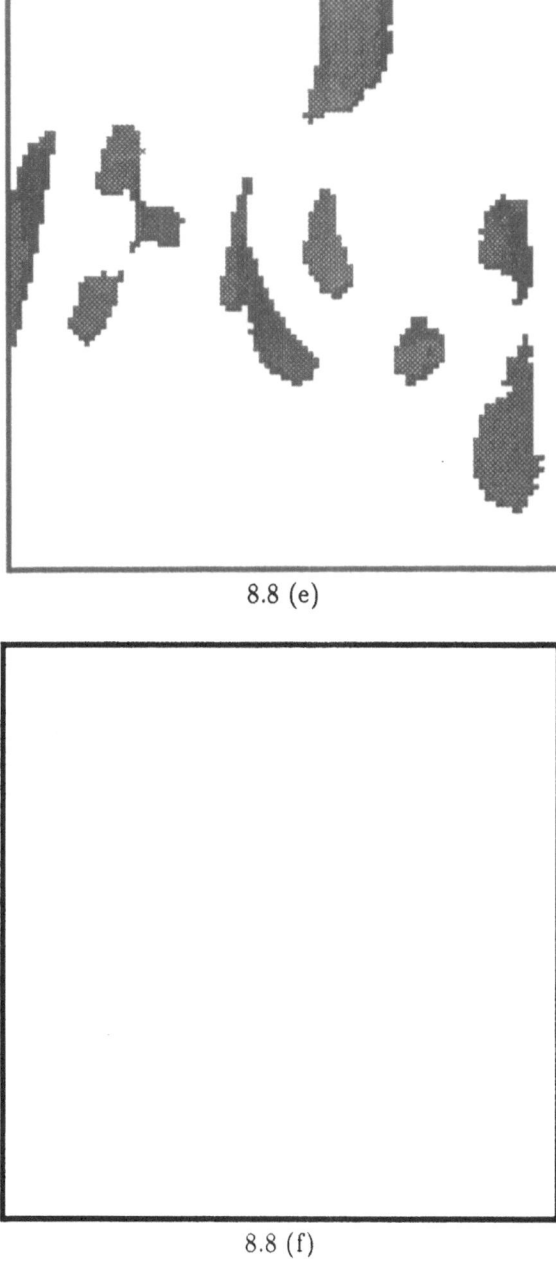

8.8 (e)

8.8 (f)

FIGURE 8.8. A subwindow of the output from the IGA algorithm for each of the sequences processed. Note that for the smaller smoothing windows and the no smoothing case, the resolution is quite good, but the results are inconsistent. For the medium size smoothing window, the resolution is a bit poorer, but the results are much more consistent, and for the large smoothing windows, the resolution is poor (or lost completely) and the accuracy begins to degrade. (a) no smoothing. (b) 2 × 2. (c) 4 × 4. (d) 8 × 8. (e) 16 × 16. (f) 32 × 32.

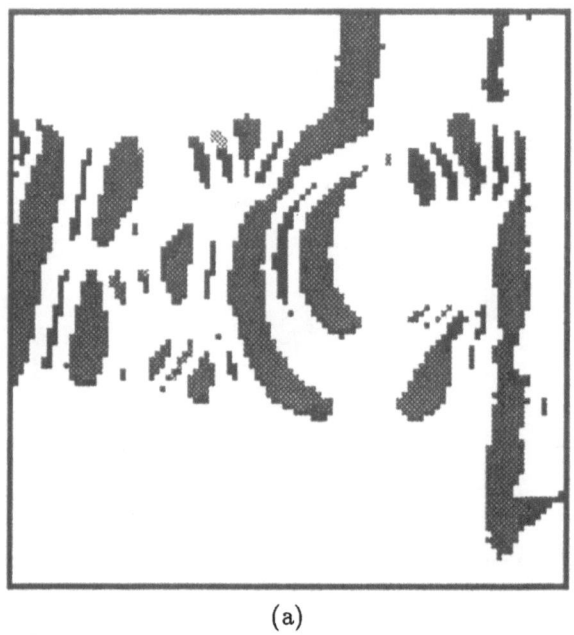

(a)

(b)

FIGURE 8.9. The output of the IGA algorithm using an 8 × 8 smoothing window (a) before smoothing is done on the output, and (b) after smoothing is done.

FIGURE 8.10. An indoor scene used to illustrate the effects of selecting different threshold (T) values on the performance of IGA.

as to include all image locations where depth information can be recovered. Clearly, since the temporal stability of the input to IGA is dependent on the size of the smoothing filter used, the threshold value selected for a given IGA-based depth recovery system will change with the size of the smoothing filter used.

Figure 8.10 shows a simple indoor scene. The corner of the room is 670cm from the camera, and the clock on the wall is 580cm from the camera (measured along the line of sight). In this example, lateral camera motion was used, with 1cm spacing between frames ($\phi = 90°$, and $dz = 1.0$). An 8 × 8 smoothing window was used, and the Cosmicar 8.5mm lens was used ($sw = 8$ and $f = 695$), and a factor of 4 subsampling was done (400 × 400 images were acquired, but 100 × 100 output was produced). Figures 8.11 (a) - (e) show the points identified by IGA as possible locations where depth information can be recovered for different threshold values, given the above scenario.

Note that in the cases where $T = 1$ and $T = 2$, the number of points indicated as possible locations where depth can be recovered is much greater than would be expected, and the number of points indicated in the case where $T = 16$ is less than would be expected. The locations indicated in the $T = 4$ and $T = 8$ cases seem, qualitatively, about what one would expect with the given input.

The performance of IGA in each of the above scenarios provides no surprises. In the cases where the threshold is low ($T = 1$ and $T = 2$) and the temporal stability of the input is, therefore, still an issue, the output of IGA is quite noisy, with several image locations being incorrectly identified as lying closer to the sensor than they actually are. Similarly, in the case where T is large ($T = 16$) and only the very strong visual depth cues (which are much less susceptible to the problems of temporal instability) are selected, the results of IGA are excellent. The locations selected are correctly identified as lying between 521 and 695cm from the camera, however the resulting depth map is quite sparse. In the case where $T = 4$, the vast majority of image locations selected for depth recovery are correctly indicated as lying between 521 and 695cm from the observer, however a few points (on the clock) are identified as lying between 348 and 521cm from the observer. For the $T = 8$ case, all the image location are correctly identified as lying between 521 and 695cm from the observer.

8.3.3   DISPARITY

In the derivation of the IGA algorithm (Chapters 4 and 5), it was shown that the IGA approach can be used to identify an arbitrary fixed disparity of size kx_p, where k is an integer and x_p is the width of one pixel, given that a certain assumption is valid. This assumption is that if a disparity of k pixels is to be detected at image location n, then the grey-level surface between pixel locations n and $n - k$ must be such that, if the intensity I perceived at location $n - k$ in the reference frame is less than the intensity perceived at location n in the reference frame, then the following must hold:

$$I_0(n - k) < I_0(m), \quad n - k < m \leq n. \tag{8.2}$$

Similarly, if $I_0(n - k) > I_0(n)$, the following must hold:

$$I_0(n - k) > I_0(m), \quad n - k < m \leq n. \tag{8.3}$$

In general, the larger the disparity to be detected (the larger k), the greater the probability that this assumption will not be valid. Given this observation, one might conclude that, for the most general case, it may be best for IGA to detect the smallest possible disparity ($k = 1$). There is, however, a tradeoff that must be considered when selecting the disparity detected by IGA. Recall that computed depth is inversely proportional to disparity (δ), which implies that the size of the depth range (u) returned by the algorithm is also inversely proportional to disparity:

$$u = (dz_i - dz_{i-1})\frac{r}{\delta} \cdot \cos \phi \tag{8.4}$$

Therefore, for a given camera spacing, the larger the disparity the finer the depth resolution.

Consider a scenario where lateral camera motion is used, with 1cm spacing between frames and a lens with focal length of 1000 pixels. If IGA is set

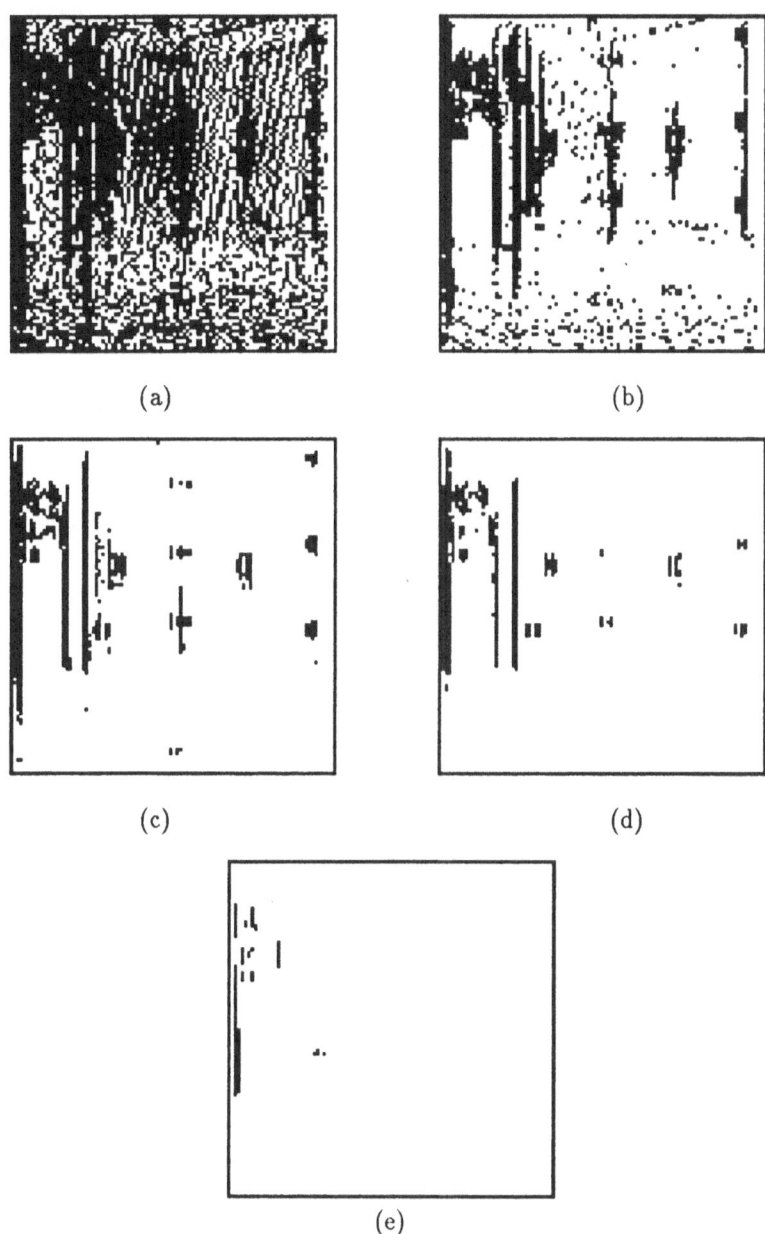

FIGURE 8.11. The points selected as possible locations where depth may be recovered for different values of T. (a) $T = 1$. (b) $T = 2$. (c) $T = 4$. (d) $T = 8$. (e) $T = 16$.

up to identify a disparity of 1 pixel, the first FDS will be located 1000cm (ten meters!) from the camera, the second 2000cm, and so on. Clearly, this may be insufficient resolution for many applications. If, however, a disparity of 10 pixels is identified, the first FDS will be located one meter from the camera, the second two meters, and so on.

It should be noted that if the above assumption about the nature of the grey-level surface is not valid, IGA must identify the object as being located *closer* to the sensor than it is actually located. That is, the temporal intensity gradient must equal or exceed the spatial intensity gradient for some sensor displacement less than that defined by the actual location of the object. This is directly analogous to the case where the spatial frequency of visual features is greater than that perceivable by the imaging array (recall Chapter 6).

Figure 8.12 shows a scene used to illustrate the tradeoffs involved in selecting the magnitude of the disparity identified by the IGA algorithm. This scene shows the CARMEL mobile platform, located 400cm from the camera (since the Cybermation platform is hexagonal, the left and right edges are actually located between 430 and 450cm from the camera and the mast for the tether is located 480cm from the camera). The white styrofoam barrier and back wall are located 525cm and 550cm from the camera, respectively. Lateral camera motion was used, with 0.5cm spacing between frames ($\phi = 90°$, and $dz = 0.5$). An 8 × 8 smoothing window was used, and only those image locations whose spatial intensity gradient was greater than 8 greylevels were considered for the depth determination process ($sw = 8$ and $T = 8$). Again, the Cosmicar 8.5mm lens was used ($f = 695$), and a factor of 4 subsampling was done (400 × 400 images were acquired, but 100 × 100 output was produced).

Figures 8.13 (a) - (f) show the output of the IGA algorithm for scenario described above. Disparities of 1, 2, 4, 8, 16, and 32 pixels were used. In the case of $\delta = 1$ (Figure 8.13 (a)), the size of the depth range was 348cm, and the algorithm correctly indicated the wall, barrier, and robot as lying between 348 and 695cm from the camera. For $\delta = 2$ (Figure 8.13 (b)), the size of the depth range was 173.75cm, and the algorithm correctly indicated the wall and barrier as lying between 521 and 695cm from the camera, and the robot was correctly identified as being located between 348 and 521cm from the camera. In the case of $\delta = 4$ (Figure 8.13 (c)), the wall and barrier were both correctly identified as being located between 521 and 605cm from the camera, and the robot was correctly identified as lying between 348 and 435cm and 435 and 521cm from the camera. Note that for the $\delta = 4$ case, one point on the robot, corresponding to a cable that is three pixels wide in the reference image, was incorrectly indicated as lying between 261 and 348cm from the camera. For the case of $\delta = 8$ (Figure 8.13 (d)), the effects of spatial frequency begin to become more evident. The majority of the image locations corresponding to the wall are identified as being located between 564 and 608cm from the camera and the barrier is indicated as

FIGURE 8.12. The scene used to illustrate the effects of different values for δ on the performance of IGA. The mobile platform, CARMEL, is located 400cm from the camera.

lying between 521 and 564cm from the camera. Similarly, the majority of the image locations corresponding to the robot were correctly identified as being located between 390 and 435cm and between 435 and 471cm from the camera. However, many image locations corresponding to visual depth cues smaller than 8 pixels in size were incorrectly identified. For example, the black border surrounding the white nameplate on CARMEL is 5 pixels wide, and was incorrectly indicated as lying closer than 400cm to the camera. Similarly, the tether, which is 4 pixels wide was incorrectly indicated as lying much closer than its actually is. As can be seen from the $\delta = 16$ and $\delta = 32$ cases (Figure 8.13 (e) and (f)), IGA is incapable of correctly identifying the location of the robot, because the spatial frequency of the features in the image is simply greater than that perceivable by sampling every 16 or 32 pixels. Note, however, that the performance of IGA does degrade as expected, with the depth values incorrectly reported for the objects in the scene are all *less than* the actual values.

8.3.4 CAMERA SPACING

In the IGA depth recovery paradigm, objects are identified as lying in a depth range that is defined by the location of the object in the image, the orientation of the camera with respect to the axis of translation, the disparity detected and the camera displacement required to induce that

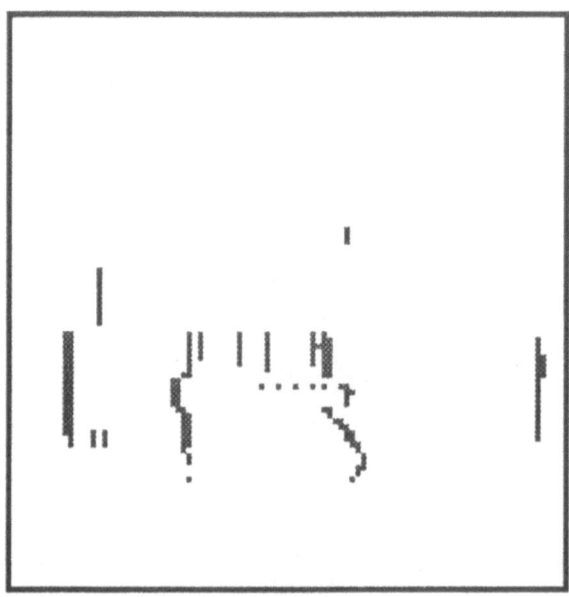

8.13 (a)

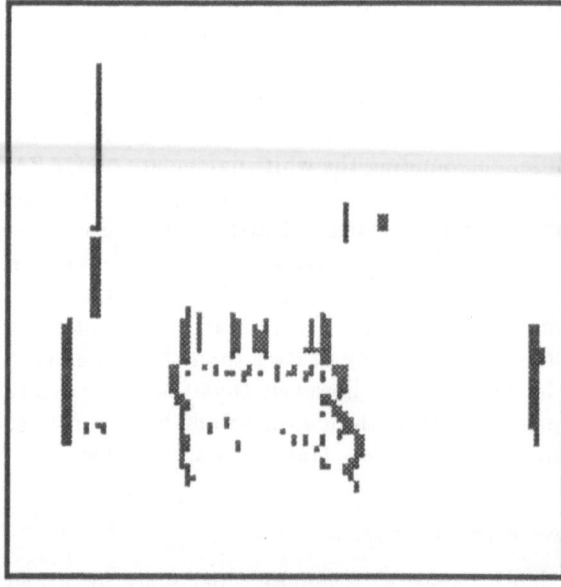

8.13 (b)

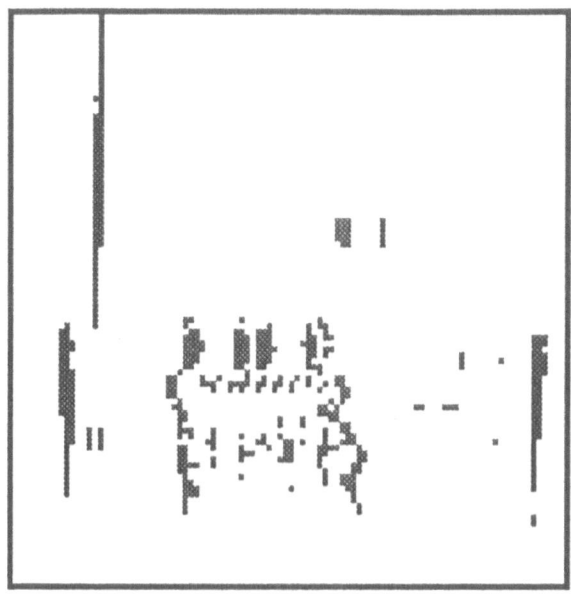

8.13 (c)

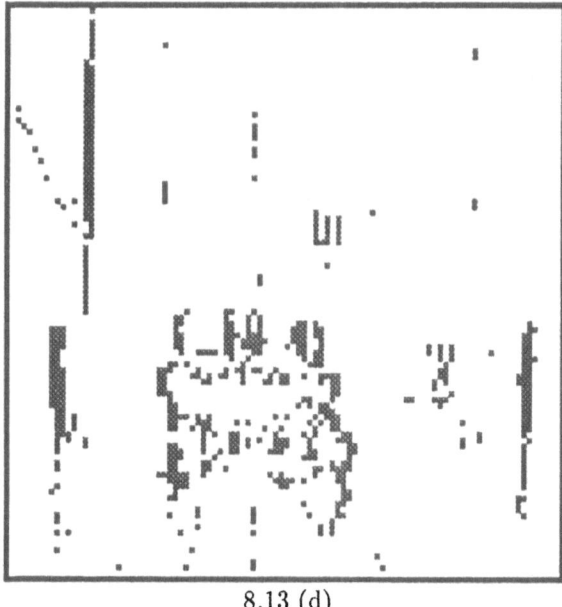

8.13 (d)

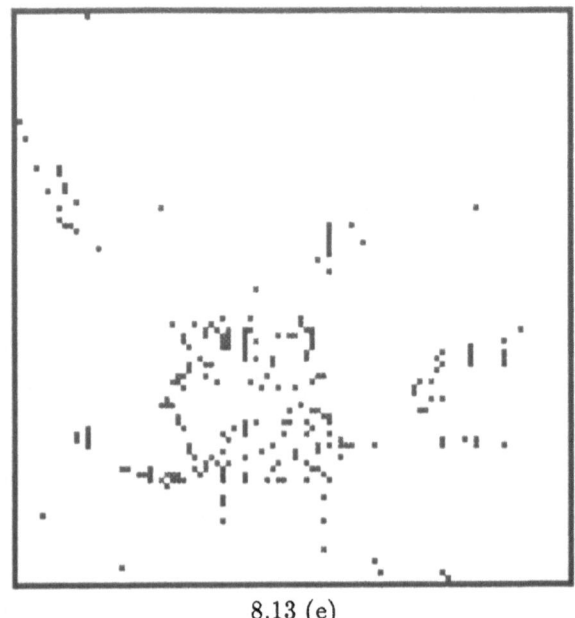

8.13 (e)

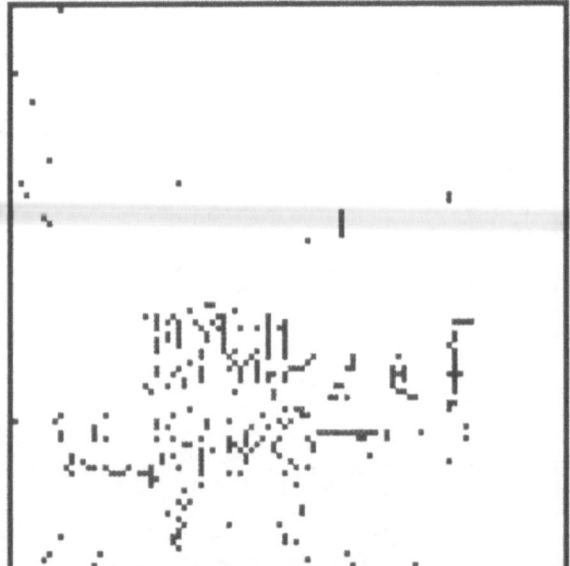

8.13 (f)

FIGURE 8.13. The output of the IGA algorithm for different values of δ. (a) $\delta = 1$. (b) $\delta = 2$. (c) $\delta = 4$. (d) $\delta = 8$. (e) $\delta = 16$. (f) $\delta = 32$.

disparity. In particular, the size of the depth range is linearly dependent on the distance the camera displaces between the last image in which the disparity was not detected and the first image in which the disparity is detected (see Equation 8.4). Therefore, to achieve fine resolution in depth, small camera displacements should be used.

Of course, there is a tradeoff between speed and resolution. Consider a simple scenario, where lateral camera motion is to be used, a disparity of ten pixels is to be detected, and a lens with a focal length of 1000 pixels is used. Suppose that in this scenario, a robot is to use IGA to navigate through a corridor that is ten meters long. If, for example, it is decided that depth resolution of 10cm is required, the camera would have to be moved in 1mm increments, implying that 100 images would have to be processed in order to determine the locations of all objects within the corridor. However, if resolution of 100cm is required (for navigation coarse resolution may be sufficient), only ten images, acquired at 1cm spacing, would have to be processed to perceive the corridor at this resolution.

Figure 8.14(a) shows a scene used to illustrate the effects of different camera spacing on the performance of IGA. This scene shows, among other things, a stack of boxes, located 7.2m from the camera, a table with computer equipment (on the left), 7.9 - 9.1m from the camera, a Denning mobile robot (in the center of the image), located 10.7m from the camera, some white obstacles (on the right), 11.3 - 12.2m from the camera, and another table with a computer workstation on it (on the right), located 13.7m from the camera. In this example, the camera was moved with an orientation of 45° with respect to the axis of translation ($\phi = 45$). A factor of 4 subsampling was done, and a disparity of 1 pixel was detected ($ss = 4$ and $\delta = 1$). An 8×8 smoothing window was used, and only those image locations whose spatial intensity gradient was greater than 8 greylevels were considered for the depth determination process ($sw = 8$ and $T = 8$). Once again, the Cosmicar 8.5mm lens was used.

Figure 8.15 shows the locations where the IGA algorithm was able to recover depth information in this scenario. Sequences were acquired using camera spacing of 2, 5, 10 and 20mm. The results are summarized in Table 8.1. Note that since lateral camera motion was not used, the size of the depth range is dependent on both the camera spacing and the location of the object in the image. This is illustrated by the difference in the size of the depth range for objects on the left hand side of the image (closer to the FOE) as compared to the size of the depth range for objects on the right hand side of the image.

It should be noted that due to the imperfect nature of CCD sensors, the resolving power of IGA is not infinite. That is, there comes a point where reducing the camera spacing results in *inaccurate* depth estimates. Consider Figure 8.16, which shows a box that is located 198cm from the camera. For this example, lateral camera motion was used and a disparity of ten pixels was detected ($\phi = 90°$ and $\delta = 10$). An 8×8 smoothing window

FIGURE 8.14. The scene used to illustrate the effects of different values for dz on the performance of IGA.

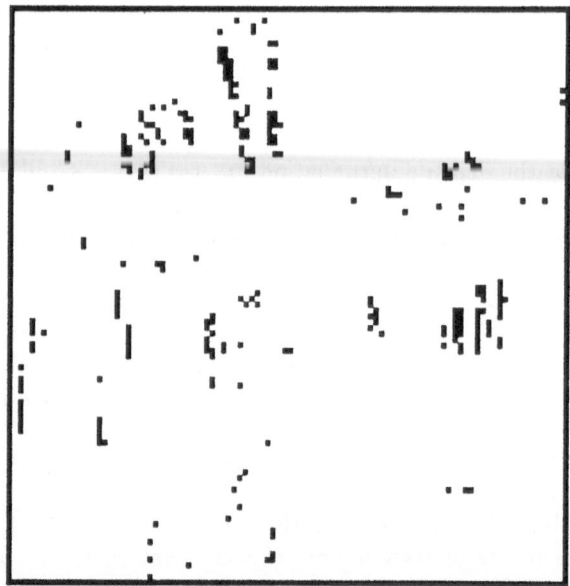

FIGURE 8.15. The locations (shown in black) where the IGA algorithm was able to recover depth information.

	Actual depth	Perceived depth for $dz =$			
	(in meters)	2mm	5mm	10mm	20mm
Boxes	7.2	4.9-5.8	4.0-5.8	4.0-7.9	0-7.9
Table 1	7.9-9.1	7.9-8.8	7.0-9.1	4.6-9.1	0-9.1
Denning	10.7	9.8-10.7	9.8-12.2	9.8-14.6	9.8-19.5
Obstacles	11.3-12.2	11.6-12.8	11.9-14.6	12.2-17.7	11.9-23.8
Table 2	13.7	13.4-14.9	13.1-16.5	10.7-15.8	10.7-21.4

TABLE 8.1. Results of IGA for different values of dz.

FIGURE 8.16. A simple laboratory scene, the box is 198cm from the camera.

was used, and only those image locations whose spatial intensity gradient was greater than 8 greylevels were considered for the depth determination process ($sw = 8$ and $T = 8$). Again, the Cosmicar 8.5mm lens was used ($f = 695$), and a factor of 4 subsampling was done ($ss = 4$).

IGA was run using 10mm spacing between frames (resulting in a depth range of 69.5cm), 5mm between frames (depth range = 34.75cm), 1mm between frames (depth range = 6.95cm) and 0.5mm between frames (depth range = 3.48cm). The results are summarized in Table 8.2. In this particular example, IGA is capable of consistently resolving the distance to the edge of the box only within a range of 6.95cm (or, more precisely, within +/- 3.5cm). This was accomplished using $dz = 1$mm. For $dz = 0.5$mm, about half of the points on the edge of the box were identified correctly as lying between 198 and 202cm from the camera, but many points on the edge of

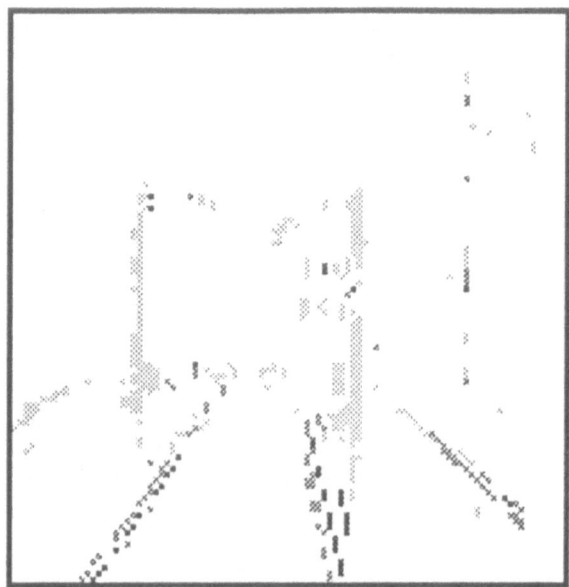

FIGURE 8.17. The output of IGA for $dz = 10mm$. Note that the errors in the center of the box are due to spatial frequency effects encountered due to the large disparity that was detected.

the box were also identified incorrectly as lying between 202 and 205cm from the camera. It should be noted that, for perfect camera motion, the resolving power is dependent on the temporal stability of the sensor, which, in turn, is dependent on the nature of the scene being imaged. Therefore, it would not be possible to infer from the above results that the performance of IGA will match the above in all possible imaging scenarios.

The output for the $dz = 1mm$ case is shown in Figure 8.17. Note that some points on the box are incorrectly indicated as lying closer to the camera than they actually due (these points appear darker than the points on the edge of the box). These errors are due to the spatial frequency effects encountered as a result of using a large disparity (see section 8.3.3).

8.3.5 ORIENTATION ANGLE

Imaging geometry plays a critical role in defining the perceptual capabilities of an IGA-based depth recovery system. This fact was discussed at considerable length in a previous chater. Therefore, the reader is referred to the discussion in Chapter 7 for a complete description and examples of the impact of imaging geometry on the perceptual capabilities of IGA.

dz (mm)	Perceived Range (cm)	Size of Range (cm)
10	139 - 209	69.5
5	174 - 209	34.8
1	195 - 202	7.0
0.5	198 - 202	3.5
	202 - 205	3.5

TABLE 8.2. Results of IGA for different values of dz.

8.3.6 Lens Focal Length

Although the focal length of the lens does not figure directly into the depth recovery equation for situations other than those involving lateral sensor motion, it is implicitly included in the measure of image distance from a point in the scene to the FOE. Recall the relationship derived in Chapter 3:

$$r = \sqrt{y^2 + f^2 \tan^2 \phi + 2fx \tan \phi + x^2} \qquad (8.5)$$

where r is the distance from a point located at image location (x, y) to the FOE, f is the focal length of the lens, and ϕ is the angle of orientation of the camera with respect to the axis of translation. Clearly, given two imaging scenarios that differ only in the focal length of the lens used, the r value corresponding to each image location will be larger for the scenario in which f is larger.

Note that the size of the field of view (ψ) is dependent on the inverse of the focal length of the lens:

$$\psi = 2 \tan^{-1}(\frac{p}{2f}) \qquad (8.6)$$

where p is the number of pixels in a scanline in the digitized image. Therefore, given the above relationship, for a given receptor array size, the amount of visual angle subtended by a pixel decreases as the focal length of the lens increases.

For a given imaging scenario, the angle of orientation of a point in the field of view with respect to the FOE will be constant, regardless of the lens used. Therefore, since the amount of visual angle subtended by a pixel decreases as the focal length increases, the distance r (measured in pixels) from a point to the FOE will increase as the focal length of the lens increases.

Both computed depth magnitude and the size of the range returned by IGA are linear functions of the image distance from the point in question to the FOE. Therefore, for an imaging scenario using fixed camera spacing, using a longer lens will result in both greater distance perceived, and a larger range of depth in which an object may lie. Consider a scenario in which the camera is moved in 1cm increments, with an orientation of 30° with respect to the axis of translation, and a disparity of four pixels is detected ($\delta = 4$). Suppose for one lens, a particular location in the field of view appears 100 pixels from the FOE. If the temporal gradient first exceeds the spatial gradient in, say, the fifth frame in the sequence, IGA will conclude that there must exist an object at that location in the field of view that lies between $4 \cdot \frac{100}{4} \cdot \cos 30$ and $5 \cdot \frac{100}{4} \cdot \cos 30$ (between 86.6 and 108.3cm) from the camera. For a longer lens, however, that same location in the field of view may appear, say, 500 pixels from the FOE. In this case, an object that appears first in the fifth frame will be identified by IGA as lying between 433 and 541cm from the camera. The object perceived with the shorter lens with a camera displacement of 5cm would be first perceived using the longer lens with a camera displacement of 1cm, and the range in which the object may lie will be between 0 and 108.3cm from the camera.

Consider the scene shown in Figures 8.18 (a) and (b). In both images, the box is 620cm from the camera, however, in Figure 8.18 (a) a 8.5mm lens is used ($f = 695$) and in Figure 8.18 (b) a 50mm lens is used ($f = 3960$). Note the significant difference in magnification (this is an excellent illustration of one way of overcoming problems associated with the spatial frequency of visual features in the scene: use a longer lens!). For this particular example, the camera was moved in 1cm increments at an angle of 30° with respect to the axis of translation. A disparity of four pixels was detected, an 8×8 smoothing window was used, and the threshold was set at 8 ($\delta = 4$, $sw = 8$, and $T = 8$). The output of IGA is shown in Figures 8.19 (a) and (b). In the case where $f = 695$ (the 8.5mm lens), the box was perceived first in the 9th frame as lying between 569 and 640cm from the camera. In the case where $f = 3960$ (the 50mm lens), the box was perceived first in the 2nd frame as lying between 445 and 910cm from the camera.

8.3.7 SUBSAMPLING

One particularly appealing property of IGA that may not be immediately apparent is that it is possible to subsample the images without affecting the maximum spatial frequency of the visual features in the scene perceivable by the algorithm. With conventional approaches, subsampling the images used necessarily implies that the maximum spatial frequency of the features perceivable is reduced. This follows directly from the fact that conventional approaches must solve the correspondence problem, and therefore, search the input for matches. Subsampling the images implies that the space searched to find matches consists of a sampling of the original images. Therefore, it must be assumed that the subsampled image is an accurate representation of the actual scene. Or, in other words, for these algorithms to work properly, the maximum spatial frequency of features in the original scene must be less than that defined by the sampling rate and the resolution of the original image.

(a)

(b)

FIGURE 8.18. Two images of a laboratory scene. (a) was acquired using a 8.5mm lens, (b) was acquired using a 50mm lens.

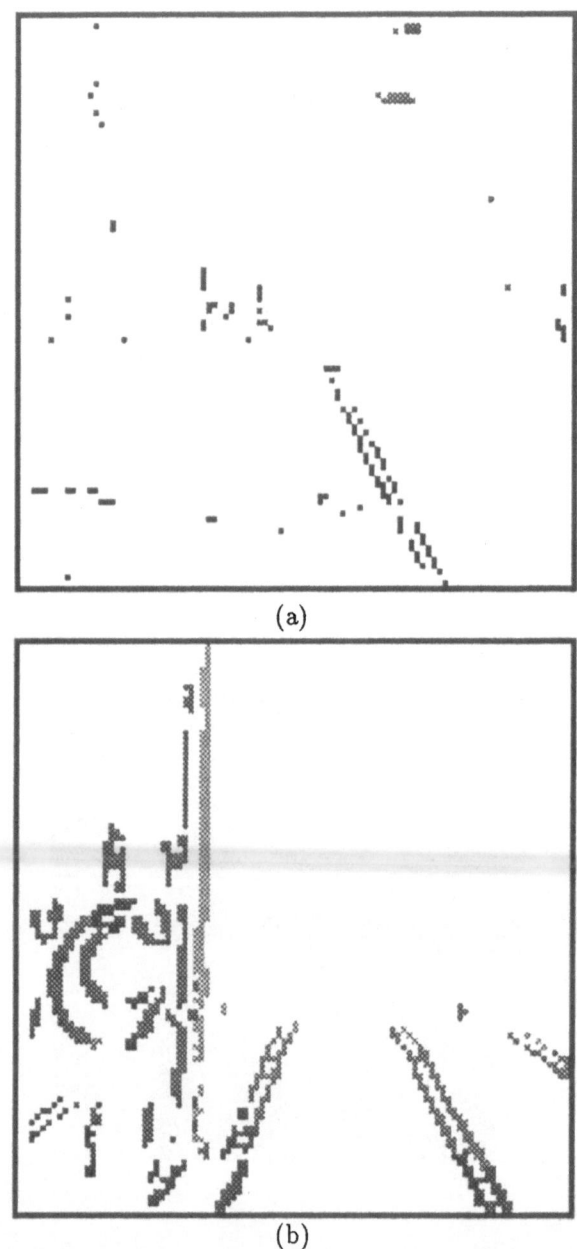

(a)

(b)

FIGURE 8.19. The output of IGA using (a) an 8.5mm lens and (b) a 50mm lens.

IGA does not suffer from the above limitation for the simple reason that IGA detects a specific type of image event. Therefore IGA does not rely directly on the assumption that maximum spatial frequency of features in the scene is less than that defined by the subsampling rate and the resolution of the original image. Assuming that the full resolution image is used to compute the spatial intensity gradient at the sampled points in the image, IGA simply monitors each "interesting" image location for the temporal intensity gradient to exceed of equal the spatial intensity gradient. No searching is done. Therefore, given an accurate spatial intensity gradient, the input can be sampled arbitrarily and still obtain results whose accuracy is dependent only on the properties of the full-resolution input. This property is significant because it implies that the execution time can be reduced (by reducing the number of image locations considered) without sacrificing accuracy.

It should be noted that subsampling should be done in a random manner, since many objects in the environment, especially man-made objects, consist of regular visual patterns (i.e. straight lines). Because of these regular patterns, if regular sampling is used, it may be the case that, if the sampling is such that one part of a visual feature is missed, the whole feature may not be perceived. For example, consider a horizontal line in a scene which lines up with an even row in the image. If only odd rows are sampled, the line may never be seen. However, if random sampling is used, the probability of such an occurrence is reduced significantly. In the above examples, regular sampling was used (for the sake of simplicity).

8.4   Outdoor Scenes

To test IGA on outdoor scenes, several experiments were performed in the parking lot behind our laboratory. Two of these experiments are described below. In the first experiment, a car is located 23.5 meters (77 feet) from the observer and, in the second experiment, the car is located 45.7 meters (150 feet) from the observer. For these experiments, a 50mm lens ($f = 3960$) was used with lateral camera motion and a factor of four subsampling was used.

Figure 8.20 shows a typical graduate student's car which lies a distance of 23.5 meters (77 feet) from the observer. An image sequence was acquired with 5mm spacing between frames, and the output of IGA is shown in Figure 8.21. Note that the algorithm correctly indicates the car as lying between 19 and 24 meters and between 24 and 29 meters from the observer. This is as expected, because the distance to the car was measured to the front bumper. Therefore, the majority of the car is greater than 23.5 meters, or between 24 and 29 meters from the observer, as reported. Note that a few locations in the output incorrectly indicated the car as lying between 9 and 14 meters from the observer. This error occurred at the right edge

FIGURE 8.20. An outdoor scene showing a car 23.5 meters (77 feet) from the observer.

of the windshield, and is due to spatial frequency effects (see Chapter 6).

Figure 8.22 again shows the same car, this time the car is located 45.7 meters (150 feet) from the observer. The output of IGA is shown in Figure 8.23. In this example, the majority of the car is identified as lying between 39 and 49 meters from the observer. A few locations on the car are identified as lying between 29 and 39 meters from the obsever and a few locations are identified as lying between 49 and 59 meters from the observer. The results of this experiment are a good illustration of the fact that the farther an object lies from the observer, the greater the potential for erroneous results due to sptial frequency effects and small errors in the camera motion. These errors are primarily due to the fact that the farther an object lies from the observer, the smaller the spatial extent of the visual depth cues on that object.

8.5   Conclusions

The experiments presented in this chapter illustrate two things: the effect of different system parameters on the performance of IGA, and the performance of IGA in several different imaging scenarios. It can be concluded from the above results that IGA performs as expected. That is, IGA provides accurate depth estimates in a wide variety of imaging scenarios. From the above experiments, it can also be concluded that proper setting

FIGURE 8.21. The output of IGA correctly indicating the car as lying between 19 and 24, and between 24 and 29 meters from the observer.

FIGURE 8.22. An outdoor scene showing a car 47.7 meters (150 feet) from the observer.

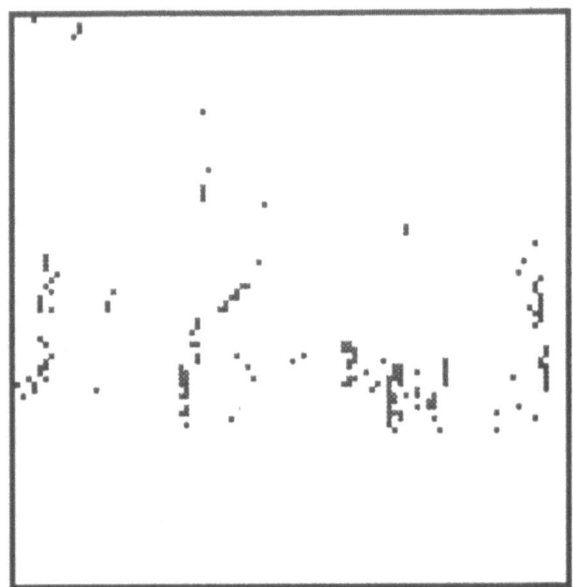

FIGURE 8.23. The output of IGA.

of the various system parameters is crucial to the successful operation of and IGA-based depth recovery system.

9

An Application: Vision-Guided Navigation Using IGA

In general, the purpose of a depth recovery algorithm is to determine the locations of objects in the field of view. However, for some tasks, such as navigation, it is equally important to determine where objects aren't located (i.e. navigable or free space). This information is inherently included in the depth estimates provided by the depth recovery algorithm: if an object at image location (i, j) is identified as lying a distance d from the observer, it is implicit that there are no objects at image location (i, j) which lie less than distance d to the observer.

In the case of IGA, the output of the algorithm can be interpreted as containing two pieces of information: a range in which an object must lie, and a range in which no objects lie. For example, if IGA indicates some image location as corresponding to an object that lies in some range $d_i - d_{i+1}$, one can also conclude that there are no objects present between the camera location and d_i (note that one can only make conclusions regarding free space at those locations in the image where IGA recovers information about filled space).

The purpose of this chapter is to demonstrate one way IGA can be used for a particular application: vision-guided navigation. In the following examples, a simple path planning algorithm is used to guide a mobile platform through a cluttered workspace using only the output of IGA. The first section briefly describes the navigation task and the efficacy of IGA for that task. The second section describes the equipment used for these experiments, the world model, and the path planning algorithm. The third section describes the results from three different scenarios.

9.1 Navigation

If a robot is to navigate from point A to point B, there must be some task for the robot to accomplish at point B: a manipulation task, an observation task, or, perhaps, a docking task. In other words, navigation is a secondary task: a task that must be performed in order for some other task to be accomplished. Because of its secondary nature, navigation should require minimal time and, perhaps more importantly, minimal (computational) effort. If an autonomous agent is to interact with its environment in an intelligent and useful manner, the limited resources on such an agent should not be dominated by those needed for a secondary task.

Note that, in order to navigate successfully though an unknown environment, the *exact* locations of objects in that environment need not be known. In fact, in most cases, coarse-grained resolution is sufficient – since it is the location of free space that is of primary concern for navigation.

Vision is the ideal mode of sensing for the navigation task – visual sensors are passive, have excellent spatial resolution, and have virtually infinite range (see Chapter 2). IGA seems ideally suited for this task, since it requires a fraction of the computational effort required by existing techniques. Additionally, IGA seems ideally suited for the navigation task because the tradeoff between resolution and computational effort is explicit. If coarse-grained resolution is desired, one could, for example, process images acquired every two centimeters instead of every one centimeter, resulting in depth estimates with half the resolution, while using half the computational effort. If one was to use a conventional binocular stereo approach, such a tradeoff would not be as straight-forward, since the time consuming search for correct correspondences must take place regardless of the imaging geometry.

9.2   Experimental Setup

This section describes the setup used to conduct the experiments on vision-guided navigation. The platform, the world model, the methodology used for integrating new information into the world model, the path planning algorithm, and the camera control strategy are all briefly described. Certainly, more complicated equipment and more robust techniques do exist, but the purpose here is merely to show the potential of IGA for such applications, not to break new ground in path planning research. The navigation algorithm assumes it is given a starting point and a destination point and works as follows:

1. move the platform to the start position.

2. orient the camera towards the destination point.

3. obtain a depth map using IGA and integrate the new information into the world model.

4. find a path through free space.

5. if destination cannot be reached, move platform to new position and/or orientation as specified by the path planner and camera control strategy. GOTO 3.

6. if goal can be reached, move platform to destination.

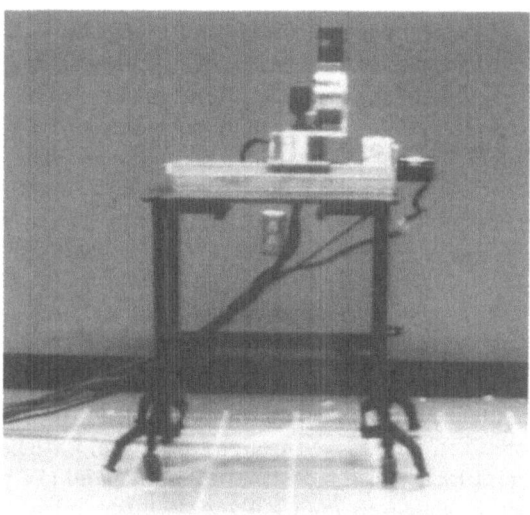

FIGURE 9.1. The Selectric mobile platform.

9.2.1 THE PLATFORM

For the purposes of these experiments, the pan/tilt/translate head used in the previous experiments (Chapter 8) was mounted on the mobile platform "Selectric." Selectric (shown in Figure 9.1) is a converted typing table with a flat top and four rolling casters that features two modes of operation: sit in one place, and BPBGS[1] locomotion. Selectric also features continuous visual motion control feedback whose accuracy is limited only by the perceptual capability of the operator.

9.2.2 THE WORLD MODEL

In these experiments the world model consists of a 64 × 64 grid, with each location corresponding to a 10cm × 10cm portion of space. An entry in the world model grid can take on one of three values: unknown, free, or filled. Before each experiment, each location on the grid is initialized to unknown.

For the sake of simplicity, it is assumed that any location in the scene perceived by IGA has infinite vertical extent. This assumption allows the world model to be two-dimensional instead of three dimensional, which, in turn, simplifies the path planning task.

---

[1] BPBGS = Be Pushed By Graduate Student

9.2.3   INTEGRATING INFORMATION

Information from different views is integrated into the world model by assuming that new information is always more reliable than old information. Therefore, for example, if the most recent output from IGA indicated voxel (i, j) as being free, the world model would be updated to indicate that voxel (i, j) corresponds to free space, regardless of what was previously concluded about that location in space. If the most recent output of IGA did not indicate whether or not voxel (i, j) corresponded to free or filled space, no change would be made to voxel (i, j).

9.2.4   PATH PLANNING

The path planning algorithm assumes that the robot moves in fixed steps of 20cm (unless the destination is less than 20cm from the current location) and that the robot can only move through free space. Paths are found using the robot's current location, the destination point, and the world model by using the algorithm shown in Figure 9.2.

Note that if the robot cannot reach its destination, and has moved since the last depth map was obtained, a new depth map is acquired. If the robot cannot reach its destination and has not moved since the last depth map was acquired, the camera is set to a new orientation using the camera control strategy, and a new depth map is obtained.

Note also that the path planning algorithm assumes that the world is simple enough so that the robot will be able to reach its destination while never moving at an orientation greater than 90 degrees with respect to the line formed by the current platform position and the destination point.

9.2.5   CAMERA CONTROL STRATEGY

In general, the camera is oriented in such a manner so as the destination point is always located in the center of the image. If no path is found using this camera orientation, the robot looks first 30° to the left, then, if still no path is found, 30° to the right (of the original orientation), then 60° left, then 60° right, and so on.

9.3   Experiments

IGA has been used to guide Selectric through several cluttered environments; the four examples presented in the following discussion are typical of the performance of this system. In the first two experiments (Four Boxes Land and The Box Jungle), lateral camera motion was used, with multiple images acquired at 5mm intervals. A disparity of 4 pixels was detected, and an 8.5mm lens with focal length equal to 695 pixels was

```
theta = orientation from current location to destination
incr = 10 degrees
foundpath = false
stuck = false
while !foundpath AND !stuck
begin
    index = 0
    foundstep = false
    while !foundstep AND index < 10
    begin
        theta1 = theta + index × incr
        theta2 = theta - index × incr
        if( free path for 20cm along orientation theta1)
                foundstep = true
                move robot 20cm along orientation theta1
        else
                if( free path for 20cm along orientation theta2)
                        foundstep = true
                        move robot 20cm along orientation theta2
                else
                        index += 1
    end
    if index = 10
        stuck = true
    else
        if current location = destination
                foundpath = true
end
```

FIGURE 9.2. The path planning algorithm.

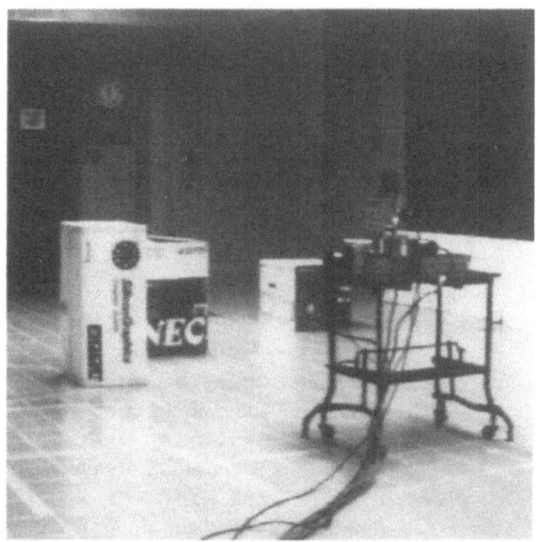

FIGURE 9.3. Four Boxes Land. Selectric must safely navigate through this dangerous maze of boxes.

used. This resulted in rather coarse resolution in depth (the size of the depth range was 86.8cm), however, the platform was able to successfully navigate through the different environments without any problem. For the third experiment (Five Box Land), the setup was identical to the previous two experiments, but axial camera motion was used and a disparity of two pixels was detected. In the fourth experiment (Virtual Barrier Navigation), lateral camera motion was used and only two images were acquired, with 5mm spacing between frames. Acquiring only two images may at first seem rather pointless, since IGA will only obtain information from a single depth range. However, this mode of operation has some interesting and desirable properties which make it quite amenable to the navigation task. These properties are discussed below.

9.3.1 FOUR BOXES LAND

Our first example takes us to Four Boxes Land, where our hero, Selectric, must safely navigate around four dangerous looking boxes. Figure 9.3 shows the setup used for this experiment, and Figure 9.4 shows a map of the world. On the map, a box on the grid corresponds to an 80 × 80cm region of space, the x's by the S and G correspond to the start and goal positions, respectively, and the shaded boxes correspond to the obstacles. The starting point is at coordinates $(250, 60)$ and the goal is located at $(340, 580)$.

From the starting point, Selectric directs its camera at an orientation

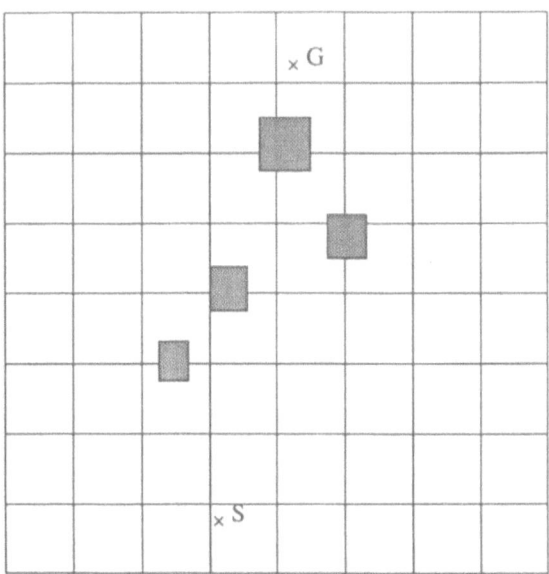

FIGURE 9.4. A map of Four Boxes Land. Each box on the grid corresponds to an 80 × 80cm region of space, and the shaded boxes correspond to obstacles. The x's by the S and G correspond to the start and goal points, respectively.

of nine degrees (with respect to the positive z axis) to look toward the destination point. Figure 9.5 shows the view seen by the robot from this position. The output of the IGA algorithm for this scene correctly indicates the three obstacles (within the tolerances of the coarse depth resolution used). The updated world model (recall that the world model is initially set to all unknown space) is shown in Figure 9.6, with white regions indicating free space, light grey regions indicating unknown space, and dark regions indicating possible filled space.

Given the depth information available from this single view, the path planner indicates that Selectric is able to move from the start point $(250, 60)$ to $(312, 398)$, which is located directly in front of the large box nearest the goal point. Figure 9.7 shows the path Selectric uses to navigate to the new location.

According to the camera control strategy, the robot must first look towards the destination point. Unfortunately, however, there is a large box that lies directly between Selectric and the goal location (see Figure 9.8). Since the box occupies the entire field of view (horizontally), no useful new information will be added to the world model using the depth map created by IGA at this location.

Since no new path is found using the depth map obtained while looking directly at the destination point, the camera control strategy indicates that the camera should be pointed 30 degrees to the left. Doing so results in the

FIGURE 9.5. The view seen by the robot at the starting point.

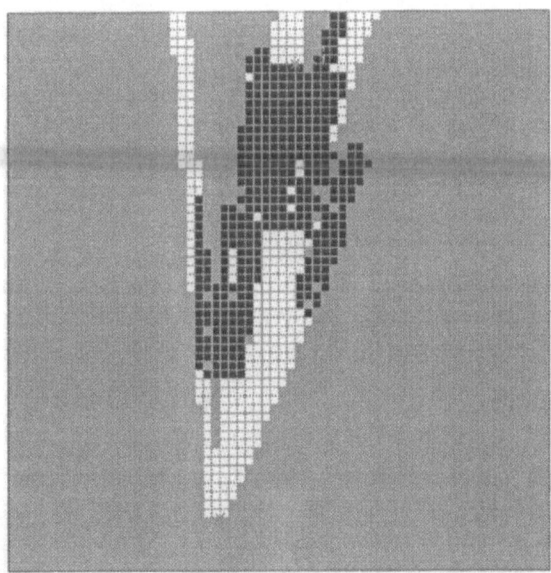

FIGURE 9.6. The world model after one revision. White indicates free space, grey indicates unknown space, and dark indicates possible filled space.

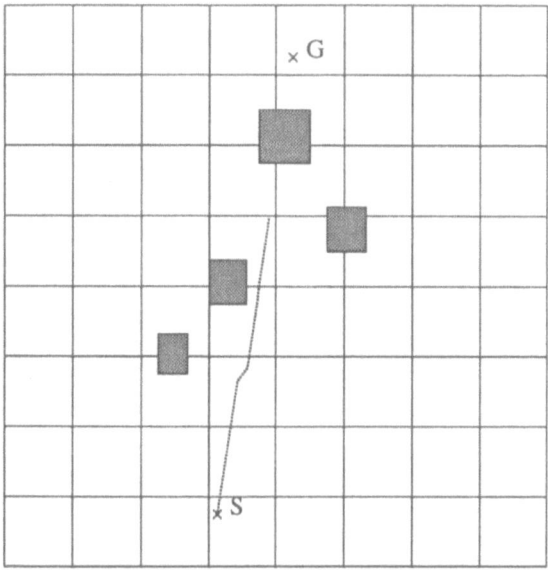

FIGURE 9.7. The path (indicated by a dotted line) extracted from the world model after one revision.

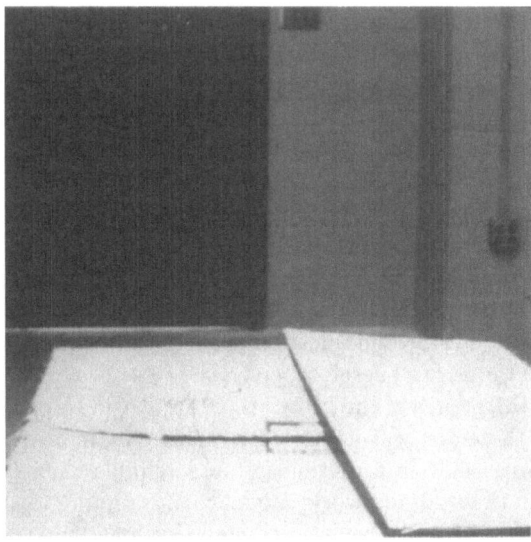

FIGURE 9.8. The view seen by the robot at location (312, 398).

FIGURE 9.9. The view seen by the robot at location (312, 398) with orientation 30 degrees to the left of the goal point.

discovery of a new navigable path that leads closer to the goal location. Figure 9.9 shows the view seen by Selectric from this new orientation and Figure 9.10 shows the world model after adding the new information from the depth map obtained with this camera orientation.

Given the new, updated world model, the path planner indicates that Selectric is now able to move from point (312, 398) to location (286, 565), which is located to the left of the goal point. Figure 9.11 shows the path Selectric uses to navigate to the new location. Note that the path selected clips the corner of one obstacle. This is not an error due to incorrect depth values produced by IGA, but rather an artifact of one of the simplifying assumptions made by the world modeling strategy. It is assumed that every object in the environment has infinite vertical extent, which, obviously is not the case. In fact, some of the obstacles are much shorter than Selectric. In particular, the obstacle in question is 50cm tall, while Selectric's camera is 86cm above the floor. Therefore, given the close proximity of the robot to the obstacle, the corner of the box did not appear in the field of view of the camera, and, therefore, was not detected. Since the information integration strategy assumes new information is more reliable than old information, the location in the updated world model corresponding to the corner of the box is identified as free space, even though this location was considered full space in the previous models.

Once Selectric reaches location (286, 565), it sees a clear path to the des-

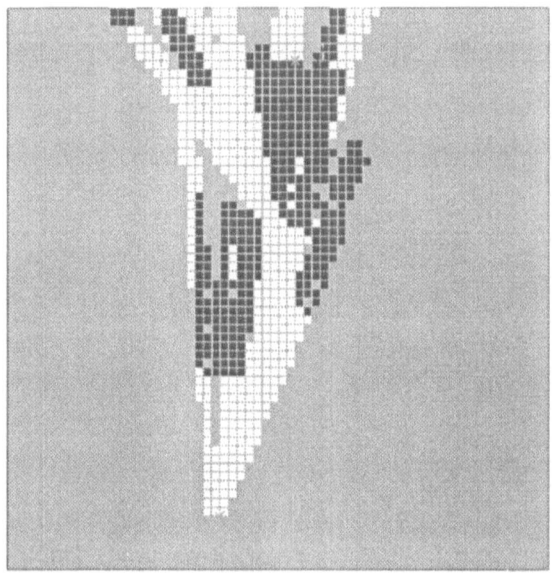

FIGURE 9.10. The world model after the new information is added.

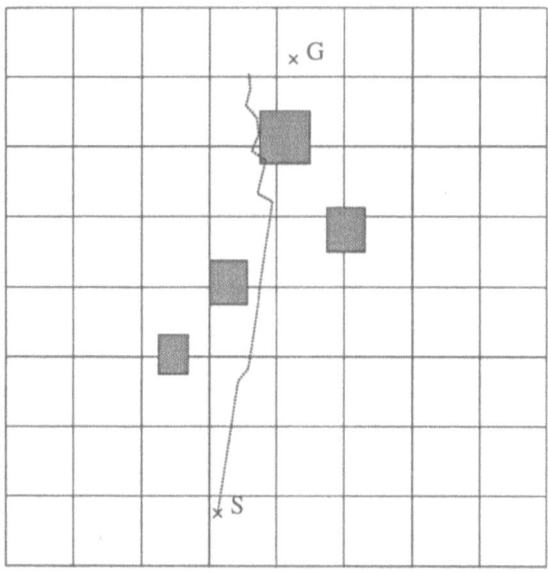

FIGURE 9.11. The path (indicated by a dotted line) extracted from the world model.

FIGURE 9.12. The view seen by the robot at location (286, 556).

tination point. Figures 9.12, 9.13 and 9.14 show the view seen by Selectric from this position, the updated world model after the information from the new depth map is added, and the complete path, respectively.

9.3.2   THE BOX JUNGLE

After successfully navigating Four Boxes Land, Selectric was catapulted into the Box Jungle, the home of nine of the meanest looking cardboard and aluminum obstacles ever seen. Figure 9.15 shows a map of this treacherous domain. The starting position is at location (360, 95) and the goal is located at (200, 580).

Note that the long slender obstacle shown near the center of the environment poses a particularly difficult situation for the vision-guided navigation task. Because this obstacle is nearly twice the height of Selectric and has uniform shading on its surface, visual depth cues exist only at the obstacle boundaries. Therefore, when the robot looks straight at this obstacle, little or no depth information is perceived.

Figures 9.16(a) and 9.17(a), respectively, show the robot's view from the starting position and the world model after the information from the first depth map is obtained. From this updated world model, the path planner determines that the robot may safely navigate to location (336, 169) (as is shown in Figure 9.18(a)).

According to the camera control strategy, Selectric is to direct its camera

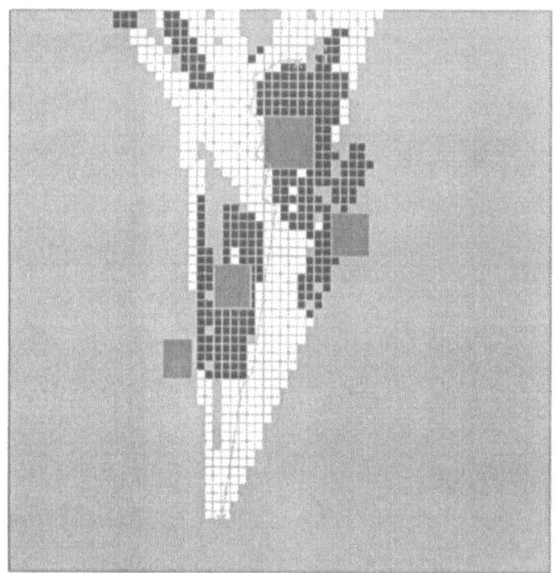

FIGURE 9.13. The final world model.

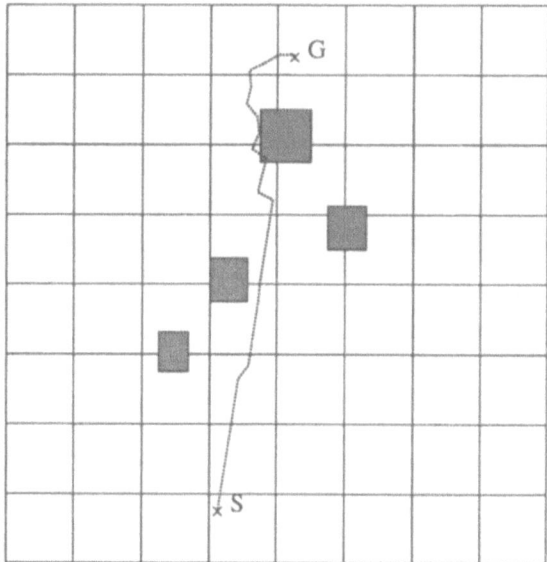

FIGURE 9.14. The complete path through 4 Boxes Land.

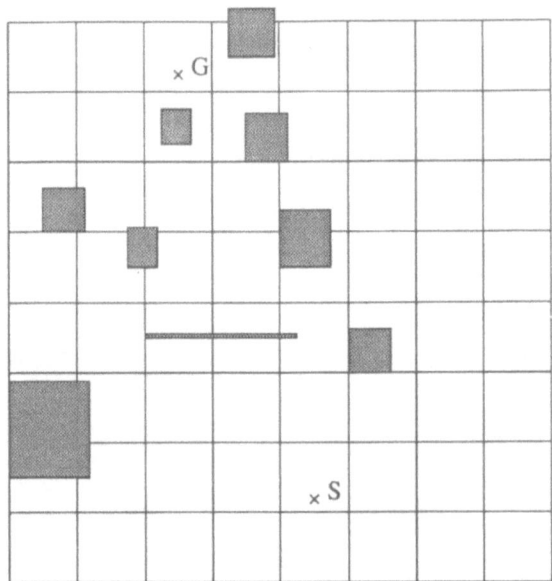

FIGURE 9.15. The Box Jungle.

towards the destination point. Unfortunately, given the large uniformly shaded obstacle directly in front of it, the robot is unable to obtain any new depth information by doing this. Therefore, also according to the camera control strategy, Selectric looks 30° to the left and sees the view shown in Figure 9.16(b). Using the depth map obtained from this orientation, the path planner indicates that the robot may safely move to location (234, 249). The world model after this new information has been added and the path found using this model are shown in Figures 9.17(b) and 9.18(b), respectively.

Figures 9.16, 9.17 and 9.18 (c) - (f) show the view seen by the robot, the world model, and the various stages of the path traveled by the robot to reach the goal. Note that, for the third view, Selectric must look 90 degrees to the left in order to find a clear path to navigate (due to the close proximity of the obstacle).

9.3.3   FIVE BOX LAND

For a change of pace, Selectric was asked to navigate though an environment containing five more abstacles, this time using axial camera motion. Figure 9.19 shows a map of the environment. The starting position is at location (290, 60) and the goal location is at (200, 520).

Figures 9.20, 9.21, and 9.22 show the robot's view, the world model as it is updated and the path selected to successfully guide Selectric though

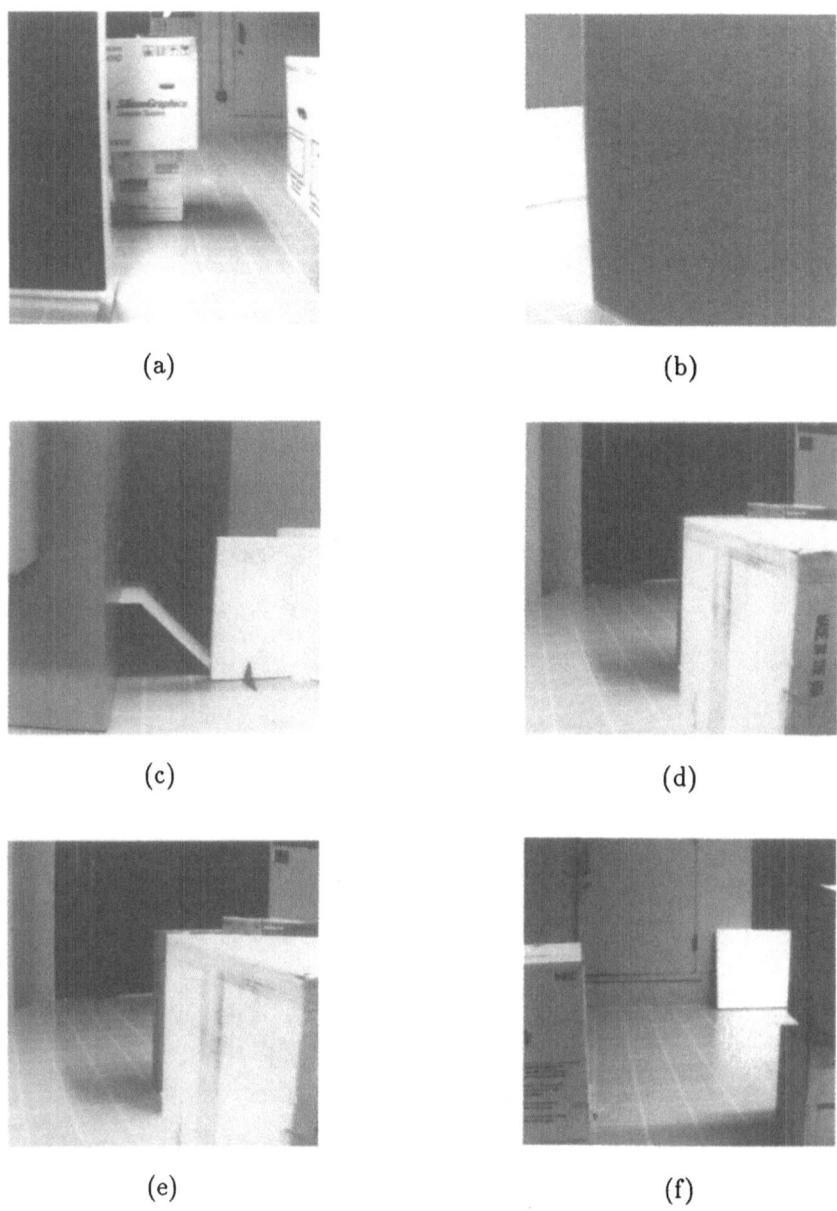

FIGURE 9.16. The Box Jungle: the robot's view from: (a) the starting position, (b) $(336, 169)$, (c) $(234, 249)$, (d) $(115, 298)$, (e) $(114, 308)$, (f) $(139, 587)$.

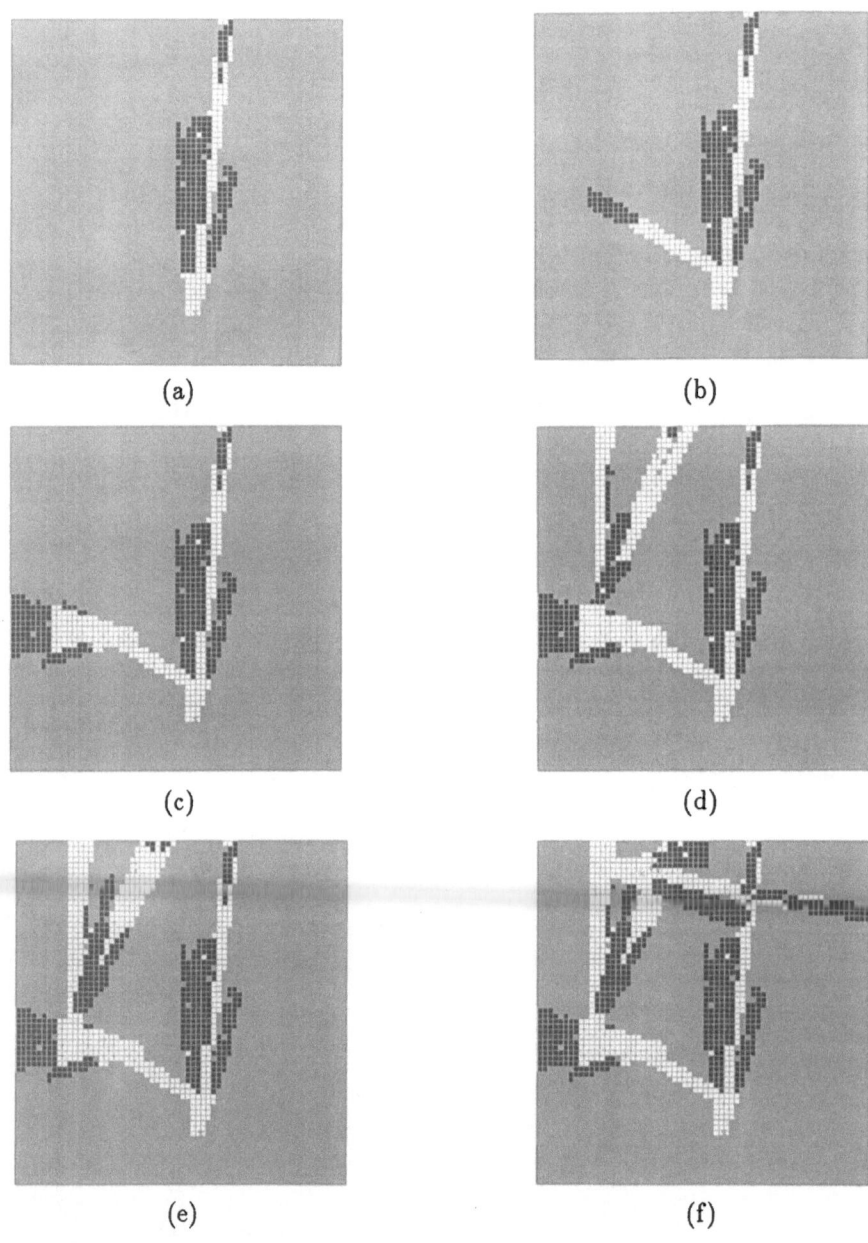

FIGURE 9.17. The Box Jungle: the world model after a depth map has been obtained from: (a) the starting position, (b) (336, 169), (c) (234, 249), (d) (115, 298), (e) (114, 308), (f) (139, 587).

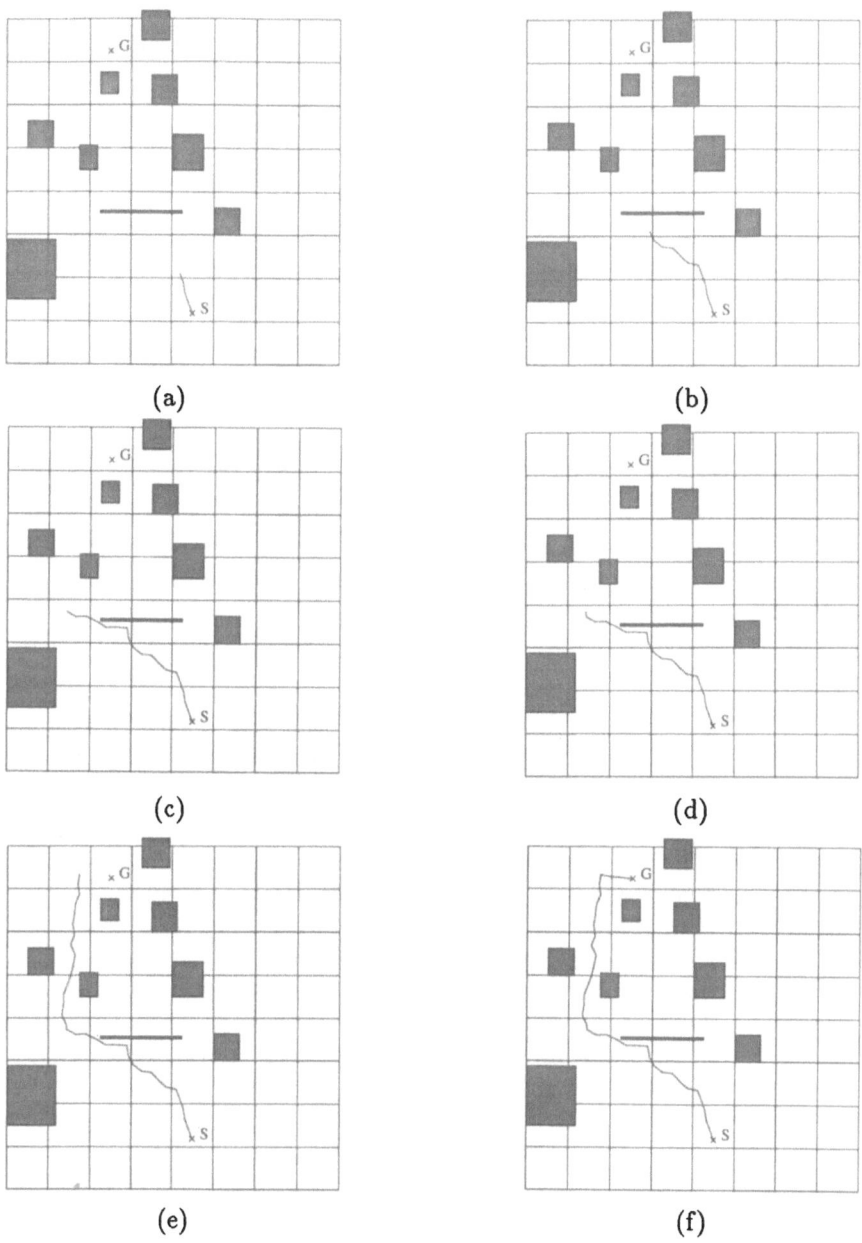

FIGURE 9.18. The Box Jungle: the path chosen after a depth map has been obtained from: (a) the starting position, (b) $(336, 169)$, (c) $(234, 249)$, (d) $(115, 298)$, (e) $(114, 308)$, (f) $(139, 587)$.

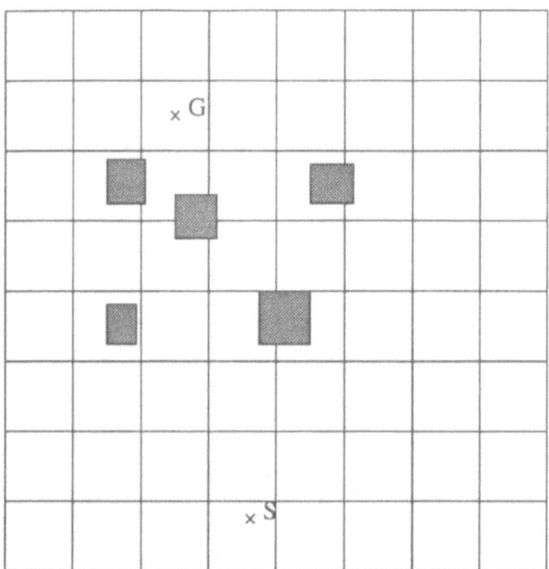

FIGURE 9.19. Five Box Land.

this environment.

9.3.4 VIRTUAL BARRIER NAVIGATION

In the fourth experiment, Selectric's perceptual capabilities were altered by limiting the number of images acquired at a given position and orientation to two. In Chapter 4, it was shown that the magnitude of the temporal intensity gradient at image locations corresponding to those objects that lie less than some distance from the camera (defined by the imaging geometry, the location in the field of view, and the fixed disparity that is identified) will be greater than or equal to the spatial intensity gradient. Therefore, since only one image (in addition to the reference image) is acquired, IGA will detect objects that lie closer than the above distance to the camera, but will not be able to isolate a range in which objects farther away must lie. Fortunately, however, it can be concluded that if an image location corresponds to a visual depth cue and the temporal intensity gradient is not greater than or equal to the spatial intensity gradient, then the object must lie *farther than* the specified distance[2] from the camera. Or, more importantly, that the region between the camera and the world location at that point in the field of view must correspond to free space.

---

[2]The specified distance being the distance defined by the imaging geometry, the location in the field of view and the fixed disparity detected.

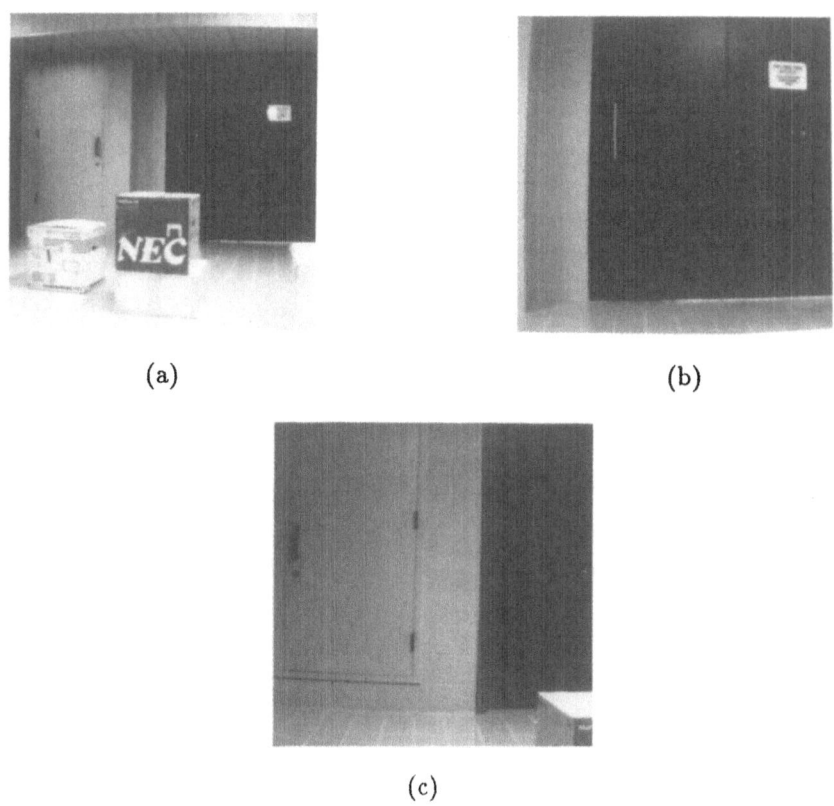

(a)   (b)

(c)

FIGURE 9.20. Five Box Land: the robot's view from: (a) the starting position,
(b) $(188, 369)$, (c) $(177, 518)$.

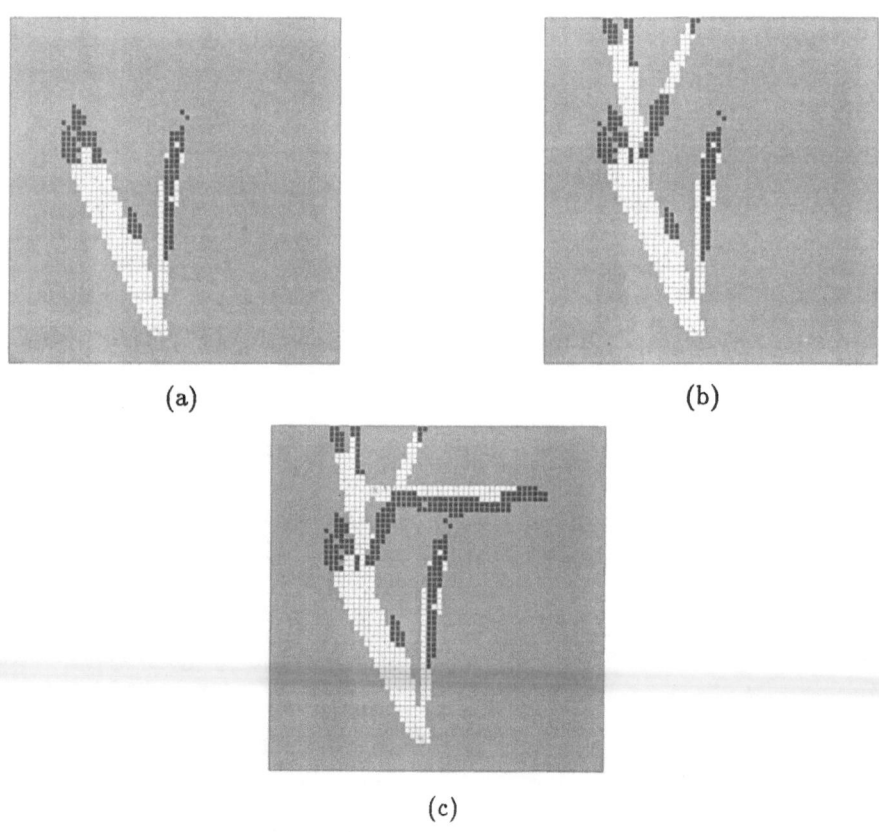

(a)

(b)

(c)

FIGURE 9.21. Five Box Land: the world model after a depth map has been obtained from: (a) the starting position, (b) $(336, 169)$, (c) $(177, 518)$.

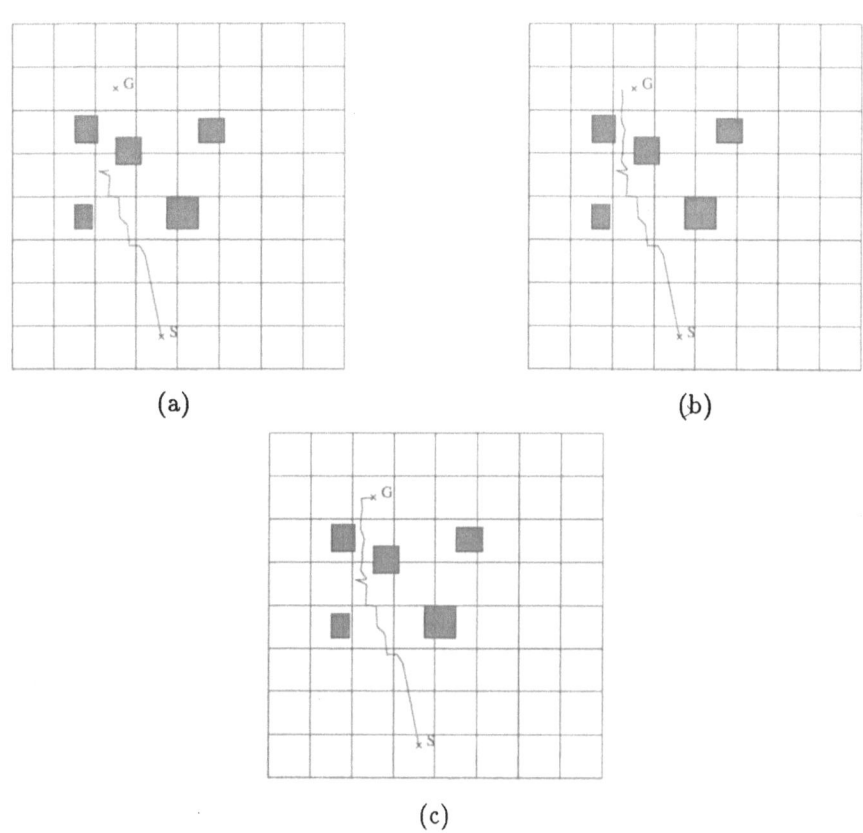

(a)

(b)

(c)

FIGURE 9.22. Five Box Land: the path chosen after a depth map has been obtained from: (a) the starting position, (b) $(336, 169)$, (c) $(177, 518)$.

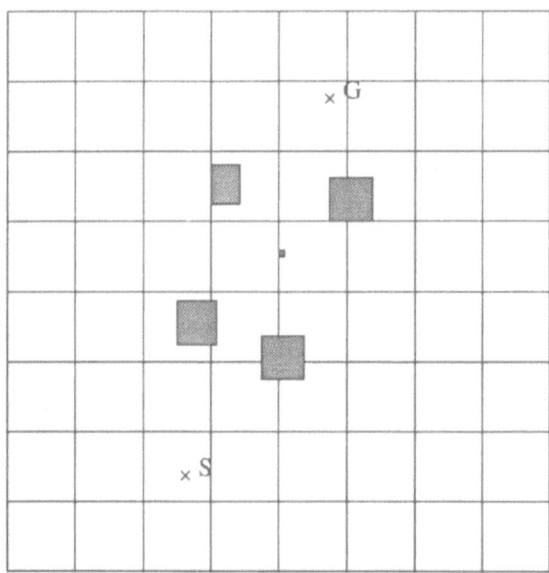

FIGURE 9.23. A map of the environment used to test virtual barrier navigation.

One way to visualize this concept is to consider the FDS defined by the imaging geometry as a virtual barrier in space (see Chapter 7). If an object penetrates or lies inside that barrier, IGA will identify a depth range in which that object must lie. However, if the object lies outside that barrier, IGA will identify the region inside the barrier as free space. This is appealing for the navigation task because (obviously) it identifies navigable space, and, more importantly, it reduces the computational effort required significantly because only two images are processed. This reduction in computational effort results in a significant speedup over the conventional implementation of IGA, which uses a sequence of several images.

Figure 9.23 shows the environment used to test the virtual barrier system. The starting point is at location $(210, 110)$ and the destination is location $(380, 540)$. Figures 9.24, 9.25 and 9.26 show the robot's view, the world model and the path selected to successfully navigate through this environment.

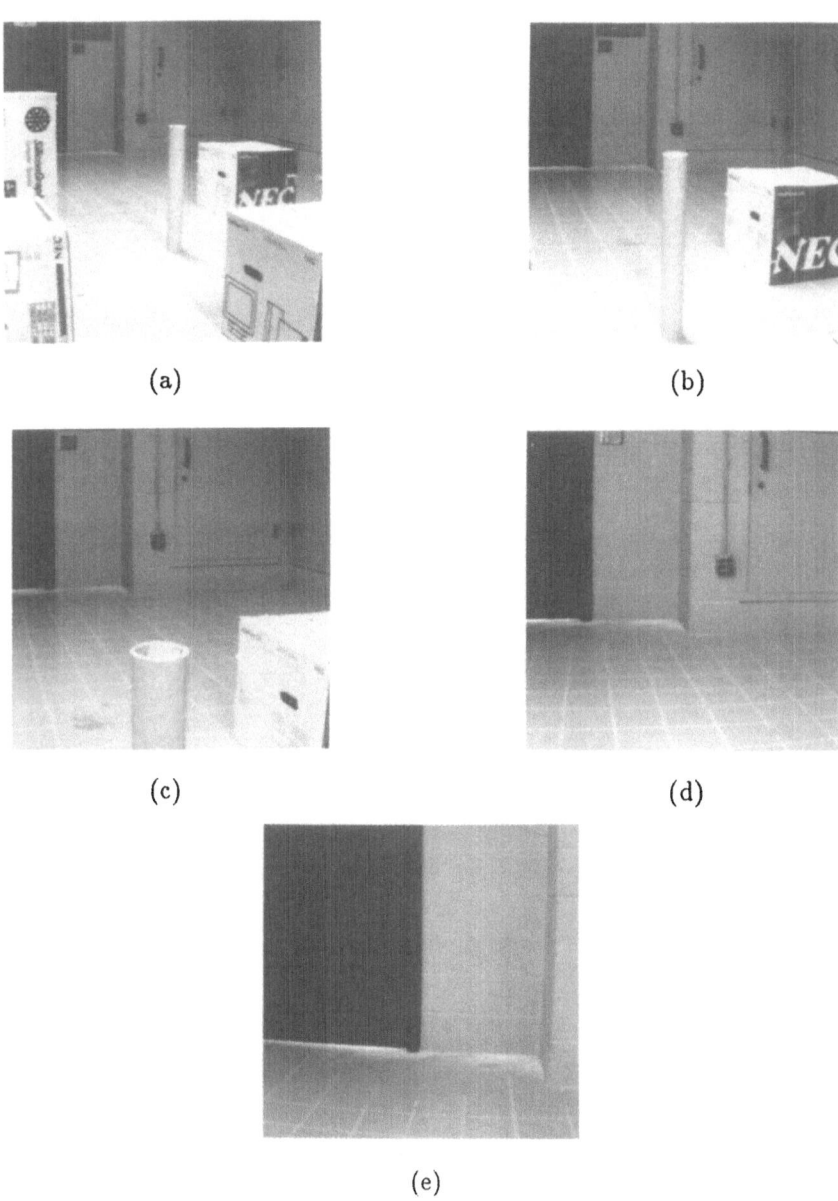

(a)   (b)

(c)   (d)

(e)

FIGURE 9.24. Virtual Barrier Navigation: the view seen by the robot from: (a) the starting position, (b) (269, 192), (c) (299, 292), (d) (352, 388), (e) (387, 489).

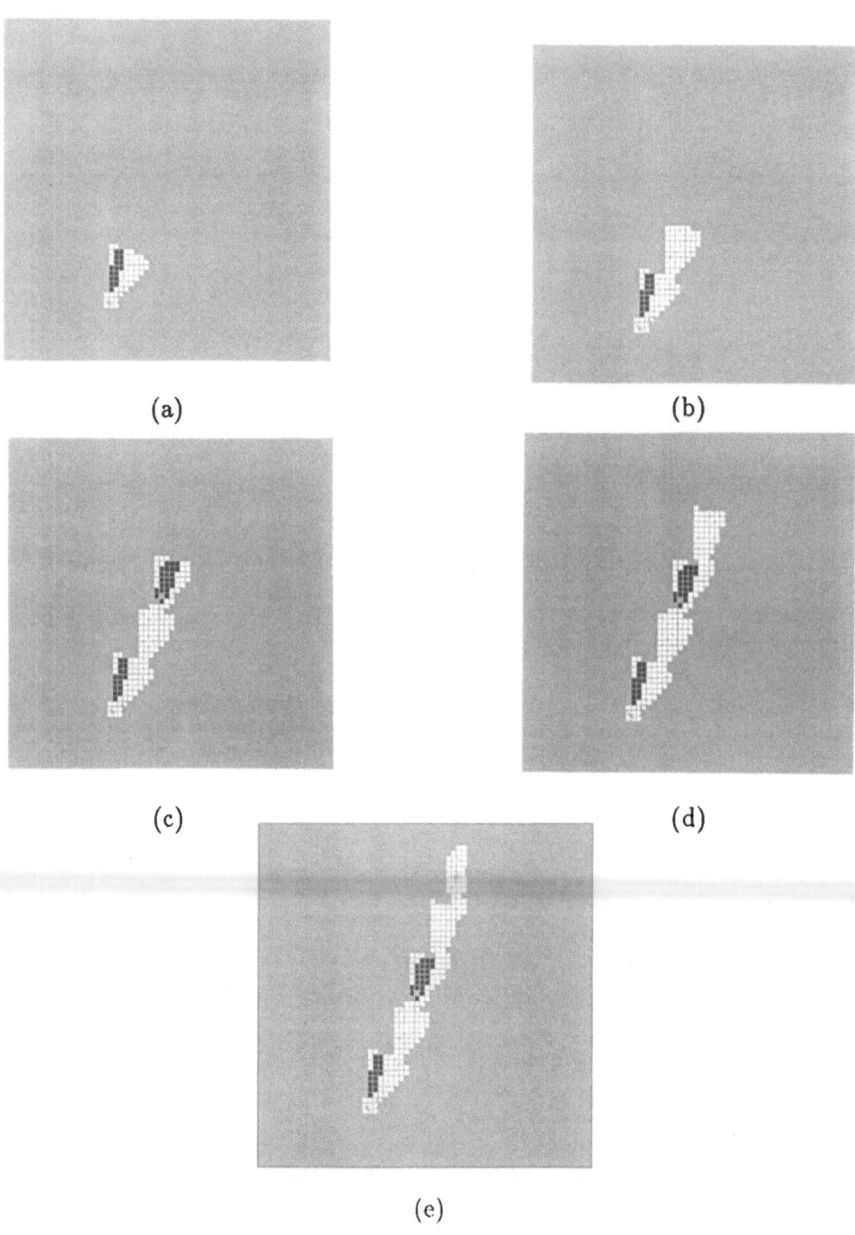

FIGURE 9.25. Virtual Barrier Navigation: the world model used by the robot after a depth map has been acquired from: (a) the starting position, (b) $(269, 192)$, (c) $(299, 292)$, (d) $(352, 388)$, (e) $(387, 489)$.

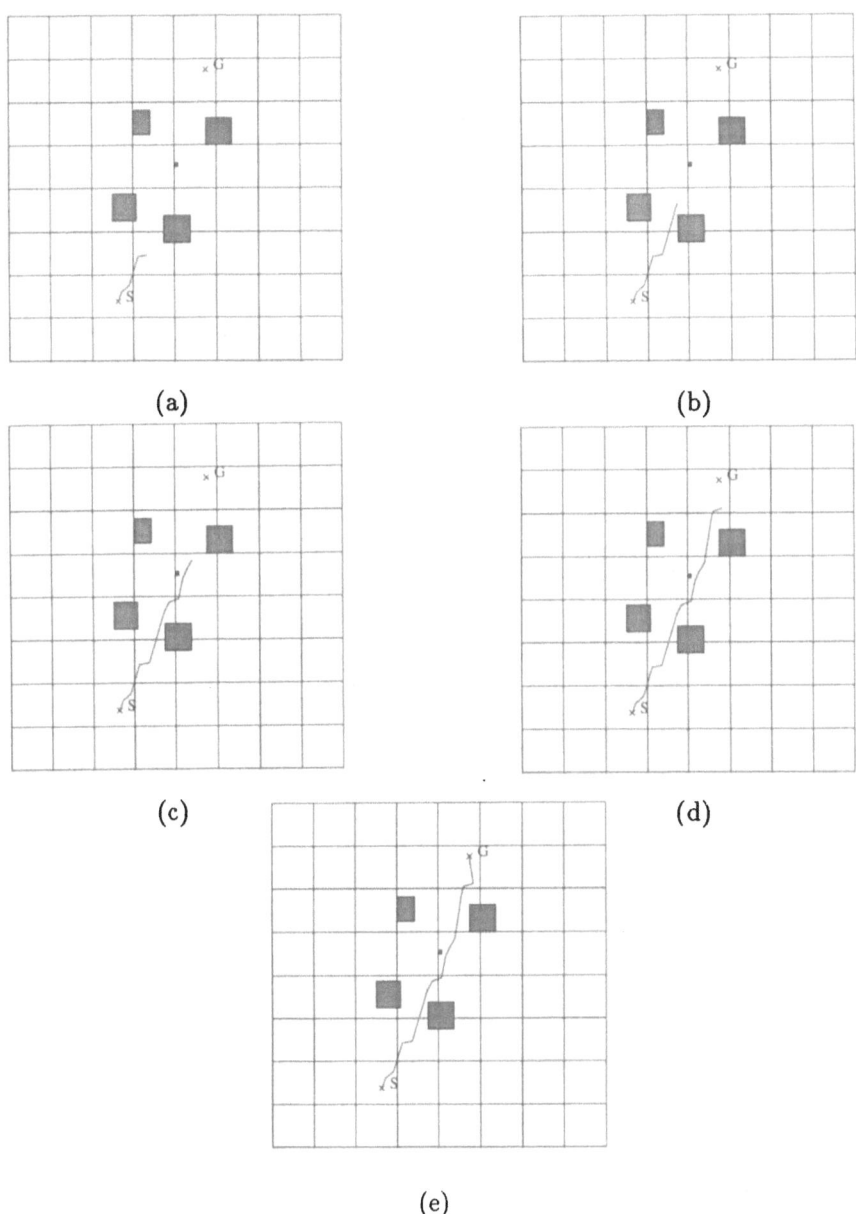

(a)   (b)

(c)   (d)

(e)

FIGURE 9.26. Virtual Barrier Navigation: the path chosen by the robot from: (a) the starting position, (b) $(269, 192)$, (c) $(299, 292)$, (d) $(352, 388)$, (e) $(387, 489)$.

10

Conclusion

An autonomous agent designed to operate in an unknown environment must be able to determine quickly and accurately the locations of objects in that environment. Vision is the most appealing mode of sensing for this task, however existing vision-based approaches have proven inappropriate for real-world applications, largely due to the excessive computational effort required by these techniques. This thesis has presented a new technique for recovering depth information from grey-scale imagery that is suitable for real-world tasks. The Intensity Gradient Analysis (IGA) algorithm produces accurate range estimates, while using a fraction of the computational effort required by existing techniques.

IGA is based on the idea that certain events can be induced in the image by simply moving the camera. These events turn out to be reliable and robust depth cues, and, perhaps more importantly, these events are detected by simply monitoring the temporal variation in image brightness. Computationally, monitoring the temporal variation in image brightness involves a single subtraction operation followed by a comparison operation (at those locations in the image where visual depth cues exist). This huge reduction in computational effort (as compared to existing techniques) is significant because there is no apparent tradeoff in accuracy involved in using the IGA approach.

Particular emphasis in this work has been placed on the issues involved in the real-world implementation of the IGA algorithm, and extensive experimentation has been done using IGA. Chapter 6 discussed the issues invovled in real-world implementation in detail, including problems with real-world sensing devices, uncertainty in the knowledge of the camera motion parameters, and problems imposed by moving objects. The experiments presented in Chapter 8 illustrate the tradeoffs involved in selecting the various parameters associated with an IGA-based depth recovery system. These experiments also illustrate the performance of IGA in several different imaging scenarios, including outdoor scenes.

The efficacy of IGA for real-world applications, such as autonomous navigation was demonstrated in Chapter 9. Several different navigation tasks were successfully completed using only the output of IGA as input to the navigation system. Both lateral camera motion and axial camera motion were used, and a simple IGA-based navigation scheme, called Virtual Barrier Navigation, was introduced.

10.1   Future Research

Work on IGA should continue along five avenues: information assimilation, the construction of dense depth maps, multi-sensor IGA, VLSI implementation, and single-sensor IGA with moving objects.

10.1.1   INFORMATION ASSIMILATION

Certainly, for an agent to reason about the environment in which it operates, it must maintain some sort of world model. For IGA to be useful to such an agent, the output of the algorithm must be assimilated into that model in an efficient and useful manner. Chapter 9 illustrates a very simple approach to this problem. For many applications, a more robust approach would have to be used.

The process of assimilating the output of IGA into a world model would be different than that associated with other techniques, due to the nature of the data. With IGA, the output consists of a range of values in which an object may lie, as opposed to the single value returned by most existing techniuqes.

10.1.2   THE CONSTRUCTION OF DENSE DEPTH MAPS

One particular limitation of any vision-based depth recovery scheme (including IGA) is the fact that depth information is recovered only at sparse locations in the image. Often, this information alone is not sufficient for an intelligent agent to reason accurately about its environment. For many systems, the reasoning task would be much easier if a dense depth map were available.

To create a dense depth map using the (sparse) output of IGA, some approach must be used to propagate the depth information throughout the image. Dense depth maps could be created by using some sort of relaxation-based approach, or by fitting a surface to the sparse depth map. Grimson [Gri81], Terzopoulos [Ter85], Poggio [Pog85], Boult and Kender [BK86], Blake and Zisserman [BZ86], and Sinha [Sin90], have studied the problem of fitting surfaces to sparse data. It remains an open issue as to how these approaches would be best adapted to deal with the range of depth values produced for each point identified by IGA, and which approach would be most appropriate for an IGA-based system.

10.1.3   MULTIPLE-SENSOR IMPLEMENTATION

A multi-sensor implementation of IGA seems particularly appealing because it effectively eliminates the issue of moving objects. Perhaps the central issue in the design of such a system is that of determining the imaging geometry. An optimal configuration would, of course, be largely application dependent. For example, if the application domain was a man-made environment, a desirable configuration may be one such that both vertical and horizontal lines in the environment provide depth cues (since man-made environments tend to have a large concentration of these features). Such a configuration could be achieved by orienting the line formed by the optical centers of the sensors so that it is not parallel to the $x - z$ or $y - z$ planes.

Ideally, a multiple sensor implementation of IGA may have variable imaging geometry. That is, both the spacing between the sensors and the relative orientation between the optical axes and the "axis of translation" (actually, the line formed by connecting the focal points of the optical systems) would be adjustable. Such a configuration would automatically adjust itself to provide the optimal imaging geometry for a given application. For example, it may be most effective to align the system so that the optical axes are perpendicular to the "axis of translation" for long-range navigation, but more appropriate to use the V-shaped imaging geometry suggested in Chapter 7 in close-range situations.

10.1.4   VLSI IMPLEMENTATION

Because of the local nature of the computations required by the algorithm, IGA seems particularly amenable to VLSI implementation. In fact, since the computational requirements of IGA are so small, it may be possible to integrate both the sensing device and the computational resources on a single chip.

A VLSI implementation of IGA would allow for the development of small, inexpensive "Depth Cameras" – real-time vision-based range sensors. The potential impact of these devices is staggering. A lightweight, microchip-sized "depth camera" would be a significant development, and would have huge ramifications in transportation, medical and aerospace applications.

10.1.5   MOVING OBJECTS AND SINGLE-SENSOR IGA

For a single-sensor implementation of IGA to operate real-world environments, it must be able to deal with moving objects. This issue was discussed in Chapter 6. It was suggested that adding a pre-processing stage that isolates moving objects may solve the problem, or the output of IGA may be combined with the output of another approach, such as binocular stereo, or laser radar.

One possible solution to the problems imposed by moving objects would be to have two (or more) sensors operating simultaneously. These sensors would be synchronized so that the first frames from their respective image sequences are acquired simultaneously. These frames could be processed as a binocular stereo pair, while the subsequent frames acquired by each sensor could be used as input into an IGA-based system. The output of IGA would then be compared to the output of the binocular stereo algorithm to detect anomalous results due to moving objects. The binocular pairs could, for example, be processed every other sequence, every fifth sequence, or every tenth sequence, depending on the nature of the environment and the speed of the image acquisition and processing hardware.

10.2   Contribution

The contribution of the work presented in this thesis is twofold. First of all, a new technique for recovering depth information from grey-scale imagery has been developed. The IGA algorithm is both fast and accurate, and has been demonstrated to have excellent potential for real-world applications, such as autonomous navigation. Conventional vision-based approaches to the depth recovery problem have proven inappropriate for such tasks, largely due to the excessive computational burden imposed by the feature extraction and correspondence steps required by these approaches. In the IGA paradigm, the depth recovery problem is reduced to monitoring the temporal variation in image brightness (a subtraction operation, followed by a comparison). Therefore, considering that vision is perhaps the most desirable sensing modality for many tasks (vision is passive, it has virtually infinite range and excellent spatial resolution), the development of IGA represents an important advancement in this area. Secondly, and perhaps more importantly, the development of IGA is a concrete example of how a complex visual task can be simplified considerably by introducing knowledge and control of sensor motion. Certainly, the computational simplicity of IGA, coupled with its proven accuracy and robustness provide a very strong argument for the efficacy of this approach. The IGA approach succeeds because is capitalizes on the strengths of robotic perceptual systems. These systems, unlike biological perceptual systems, do not have the benefit of millions of years of evolutionary fine tuning. Therefore, the emphasis in the development of working robotic systems must be on the (few) strengths inherent to the tools available. In the case of IGA, these strengths are the ability of computers and CCD devices to accurately acquire and process quantitative (as opposed to qualitative) information and the ability to know and control camera motion precisely.

Bibliography

[ABG89]   N. Alvertos, D. Brzakovic, and R. C. Gonzalez. Stereo camera modeling and image matching for 3-d machine vision. *IEEE Trans. Pattern Analysis and Machine Intelligence*, 11(9):897–815, 1989.

[AWB88]   J. Aloimonos, I. Weiss, and A. Bandyopadhyay. Active vision. *Int. J. Computer Vision*, 1:333–356, 1988.

[Baj88]   R. Bajcsy. Active perception. *Proc. IEEE*, 76(8):996–1005, 1988.

[Bal87]   D. H. Ballard. Eye movements and spatial cognition. Technical Report TR 218, University of Rochester, Department of Computer Science, 1987.

[BB82]   D. H. Ballard and C. M. Brown. *Computer Vision*. Prentice Hall, Englewood Cliffs, N.J., 1982.

[BB88]   H. H. Baker and R. C. Bolles. Generalizing epipolar-plane image analysis on the spatiotemporal surface. In *Proc. CVPR-88.*, pages 2–10, 1988.

[BBM87]   R. C. Bolles, H. H. Baker, and D. II. Marimont. Epipolar-plane image analysis: An approach to determining structure from motion. *Int. J. Computer Vision*, 1:7–55, 1987.

[Bes88]   P. J. Besl. Active optical range imaging sensors. *Machine Vision Applications*, 1, 1988. see also *Advances in Machine Vision: Architectures and Applications* (J. Sanz, Ed.), Springer-Verlag, New York; (2) Range imaging sensors. Research Report GMR-6090, General Motors Research Laboratories, Warren, Mich.

[BF82]   S. T. Barnard and M. A. Fischler. Computational stereo. *Computing Surveys*, 14(4):553–572, 1982.

[BK86]   T. Boult and J. R. Kender. Visual surface reconstruction using sparse depth data. In *Proc. Computer Vision Pattern Recognition Conf.*, pages 68–76, Miami, Fla, 1986. IEEE-CS.

[BK88]   K. L. Boyer and A. C. Kak. Structural stereopsis in 3-*D* vision. *IEEE Trans. Pattern Analysis and Machine Intelligence*, PAMI-10(2):144–166, 1988.

[BL80]   J. D. E. Beynon and D. R. Lamb, editors. *Charge-Coupled Devices and Their Applications*. McGraw-Hill Book Company (UK) Limited, London, 1980.

[BT80]   S. T. Barnard and W. Thompson. Disparity analysis of images. *IEEE Trans. Pattern Analysis and Machine Intelligence*, PAMI-2(4), 1980.

[BZ86]   A. Blake and A. Zisserman. Invariant surface reconstruction using weak continuity constraints. In *Proc. Conf. Computer Vision and Pattern Recognition*, 1986.

[CH82]   T. S. Collet and L. I. K. Harkness. Depth vision in animals. In D. J. Ingle, M. A. Goodale, and R. J. W. Mainsfield, editors, *Analysis of Visual Behavior*. MIT Press, 1982.

[Cle87]   R. Clement. Line correspondence in binocular vision. *Perception*, 16:193–199, 1987.

[DW88]   T. Darrel and K. Wohn. Pyramid based depth from focus. In *Proc. CVPR 1988*, pages 504–509, 1988.

[Elf87]   A. Elfes. Sonar-Based Real-World Mapping and Navigation. *IEEE J. Robotics and Automation*, 3, 1987.

[Gib79]   J. Gibson. *The Ecological Approach to Visual Perception*. Houghton-Mifflin, Boston, 1979.

[Gol84]   E. B. Goldstein. *Sensation and Perception*. Wadsworth Publishing Co., 1984.

[Gri81]   W. E. L. Grimson. A computer implementation of a theory of human stereo vision. *Phil. Trans. R. Soc. Lond.* B, 292:217–253, 1981.

[Gro87]   P. Grossmann. Depth form focus. *Pattern Recognition Letters*, 5:63–69, 1987.

[GY88]   N. C. Griswold and C. P. Yeh. A new stereo vision model based upon the binocular fusion concept. *Comp. Vision, Graph. and Image Proc.*, 41:153–171, 1988.

[HAH90]   M. Herman, J. S. Albus, and T.-H. Hong. Real-time vision for autonomous and teleoperated control of unmanned vehicles. In A. Sood, editor, *Active Perception and Robot Vision*. Springer-Verlag, Heidelberg, 1990.

[Hee87]   D. J. Heeger. Model for the extraction of image flow. *J. Opt. Soc. Am. A*, 4(8):1455–1471, 1987.

[HNR84]   Y. Z. Hsu, H. H. Nagel, and G. Rekers. New likelihood test methods for chagne detection in image sequences. *Comp. Vision, Graph. and Image Proc.*, 26:73–106, 1984.

[Hor86]   B. K. P. Horn. *Robot Vision*. MIT Press, Cambridge, MA, 1986.

[HS81]   B. K. P. Horn and B. G. Schunck. Determining optical flow. *Artificial Intelligence*, 17:185–203, 1981.

[IMO84]   H. Itoh, A. Miyauchi, and S. Ozawa. Distance measuring method using only simple vision constrained for moving robots. In *Proc. Seventh Int. Conf. Pattern Recognition*, pages 192–195, Montreal, 1984.

[JBO87]   R. Jain, S. Bartlett, and N. O'Brien. Motion stereo using ego-motion complex logarithmic mapping. *IEEE Trans. Pattern Analysis and Machine Intelligence*, PAMI-9:356–369, 1987.

[JH82]   R. Jain and S. Haynes. Imprecision in computer vision. *IEEE Computer*, pages 39–48, August 1982.

[JMN77]   R. Jain, D. Militzer, and H. H. Nagel. Separating non-stationary from stationary scene components in a sequence of real world TV-images. In *Proc. IJCAI 1977*, pages 612–618, 1977.

[KK88]   E. Krotkov and R. Kories. Adaptive control of cooperating sensors: Focus and stereo ranging with an agile camera system. In *Proc. IEEE Conf. on Robotics and Automation*, pages 548–553, 1988.

[KTB87]   J. K. Kearney, W. B. Thompson, and D. L. Boley. Optical flow estimation: An error analysis of gradient-based methods with local optimization. *IEEE Trans. Pattern Analysis and Machine Intelligence*, 9(2):229–244, 1987.

[Lee76]   D. N. Lee. A theory of visual control of braking based on information about time-to-collision. *Perception*, 5:473–459, 1976.

[LJ89]   S.-P. Liou and R. Jain. Motion detection in spatio-temporal space. *Comp. Vision, Graph. and Image Proc.*, 45:227–250, 1989.

[LK85]   B. D. Lucas and T. Kanade. Optical navigation by the method of differences. In *Proc. IJCAI 85*, pages 981–983, 1985.

[Luc85]   B. D. Lucas. Generalized Image Matching by the Method of Differences. Technical Report CMU-CS-85-160, Carnegie-Mellon University, 1985.

[ME85]   H. P. Moravec and A. Elfes. High-Resolution Maps from Wide Angle Sonar. In *Proc. IEEE International Conference on Robotics and Automation*, St. Louis, 1985.

[MF81]   J. E. W. Mayhew and J. P. Frisby. Psychophysical and computational studies towards a theory of human stereopsis. *Artificial Intelligence*, 17:349–385, 1981.

[Mor81] H. P. Moravec. *Robot Rover Visual Navigation*. UMI Research Press, Ann Arbor, MI, 1981.

[MP79]   D. Marr and T. Poggio. A computational theory of human stereo vision. *Proc. R. Soc. Lond.* B, 204:3201–328, 1979.

[MSK89] L. Matthies, R. Szeliski, and T. Kanade. Kalman Filter-Based Algorithms for Estimating Depth from Image Sequences. *Int. J. Computer Vision*, 3, 1989.

[Nei76] U. Neisser. *Cognition and Reality*. W. H. Freeman and Company, 1976.

[Nev76] R. Nevatia. Depth measurement by motion stereo. *Comp. Graph. and Image Proc.*, 5:203–214, 1976.

[NH86]   S. Negahdaripour and B. K. P. Horn. Direct passive navigation: Analytical solution for planes. In *Proc. Int. Joint. Conf. Robotics and Automation*, pages 1157–1163, 1986.

[NH87]   S. Negahdaripour and B. K. P. Horn. Direct passive navigation. *IEEE Trans. Pattern Analysis and Machine Intelligence*, 9(1):168–176, 1987.

[Nis84] H. K. Nishihara. PRISM: A practical real-time imaging stereo matcher. *Optical Engineering*, 23:536–545, 1984.

[OJ84]   N. O'Brien and R. Jain. Axial motion stereo. In *Proc. Workshop Computer Vision*, pages 88–92, 1984.

[OS75]   A. V. Oppenheim and R. W. Shafer. *Digital Signal Processing*. Prentice-Hall, Englewood Cliffs, NJ, 1975.

[Pee80] P. Z. Peebles. *Probability, Random Variables and Random Signal Principles*. McGraw-Hill, 1980.

[Pen87] A. P. Pentland. A new sense for depth of field. *IEEE Trans. Pattern Analysis and Machine Intelligence*, 9(4):523–531, 1987.

[Pog85] T. Poggio. Early vision: From computational structure to algorithms and parallel hardware. *Comp. Vision, Graph. and Image Proc.*, 31:139–155, 1985.

[Sin90]   S. S. Sinha. *Discontinuity Preserving Surface Reconstruction in Vision Processing*. PhD thesis, University of Michigan, 1990.

[SJ89]   K. Skifstad and R. Jain. Illumination independent change detection for real-world image sequences. *Comp. Vision, Graph. and Image Proc.*, 46:387–399, 1989.

[ST90]   G. Sandini and M. Tistarelli. Active tracking strategy for monocular depth inference over mulitple frames. *IEEE Trans. Pattern Analysis and Machine Intelligence*, 12(1):13–27, 1990.

[Sub89]   M. Subbarao. Efficient depth recovery through inverse optics. In H. Freeman, editor, *Machine Vision for Inspection and Measurement*. Academic Press, 1989.

[Ter85]   D. Terzopoulos. Computing visible surface reconstruction. Memo 800, Artificial Intelligence Laboratory, M. I. T., 1985.

[TH87]   T. Tsukiyama and T. S. Huang. Motion stereo for navigation of autonomous vehicles in man-made environments. *Pattern Recognition*, 20(1):105–113, 1987.

[TK88]   S. Tanaka and A. C. Kak. A rule-based approach to binocular stereopsis. Technical Report TR-EE-88-33, School of Electrical Engineering, Purdue University, 1988.

[Tsa83]   R. Tsai. Multiframe image point matching and 3-D surface reconstruction. *IEEE Trans. Pattern Analysis and Machine Intelligence*, PAMI-5(2):159–173, 1983.

[Vel]   Velmex, Inc., E. Bloomfield, NY. *Unislide motor driver assemblies and controls*. Catalog M-88.

[VT89]   D. Vernon and M. Tistarelli. Using camera motion to estimate range for robotic parts manipulation. Submitted to IEEE Trans. on Robotics and Automation, 1989.

[WA85]   A. B. Watson and A. J. Ahumada. Model of human visual-motion sensing. *J. Opt. Soc. Am. A*, 2(2):322–342, 1985.

[WM88]   T. E. Weymouth and S. Moezzi. Wide base-line dynamic stereo: Approximation and refinement. In *Proceedings CVPR*, pages 183–88, 1988.

[XTA87]   G. Xu, S. Tsuji, and M. Asada. A motion stereo method based on coarse-to-fine control strategy. *IEEE Trans. Pattern Analysis and Machine Intelligence*, 9(2):332–336, March 1987.

[YKK86]   M. Yachida, Y. Kitamura, and M. Kimachi. Trinocular vision: New approach for correspondence problem. In *Proc. 8th ICPR*, pages 1041–1044, 1986.

Index